Leckie
the education publisher
for Scotland

C000277240

Higher
GEOGRAPHY
For SQA 2019 and beyond

Course Notes
Fiona Williamson and
Sheena Williamson

ISBN 9780008383480

Published by Leckie & Leckie Ltd
An imprint of HarperCollins Publishers
Westerhill Road, Bishopbriggs, Glasgow, G64 2QT
T: 0844 576 8126   F: 0844 576 8131
leckiescotland@harpercollins.co.uk
www.leckiescotland.co.uk

Publisher: Sarah Mitchell
Project manager: Fiona Watson and Gillian Bowman

Special thanks to
Louise Robb (proofread)
Jouve (layout and illustration)

Printed in Italy by GRAFICA VENETA S.p.A.

A CIP Catalogue record for this book is available from the British Library.

Acknowledgements

Fig 1.25 © JAY DIRECTO/AFP/Getty Images; Fig 1.31 © Mamadou Toure BEHAN/AFP/GettyImages; Fig 2.6 © peresanz / Shutterstock.com; Fig 2.11 © 365_visuals / Shutterstock.com; Fig 2.19 © Sarah Ann Loreth / Aurora Photos / Getty Images; Fig 3.2 Till Niermann / licensed under the Creative Commons Attribution-Share Alike 3.0 Unported license; Fig 3.4 Walter Siegmund / licensed under the Creative Commons Attribution-Share Alike 3.0 Unported license; Fig 3.5 Silence-is-infinite / licensed under the Creative Commons Attribution-Share Alike 4.0 International license; Fig 3.9 Silence-is-Infinite / licensed under the Creative Commons Attribution-Share Alike 4.0 International license; Fig 3.17 Hannah Calkin / licensed under a Creative Commons Attribution Share-Alike 3.0 License; Fig 3.18 Mick Knapton / licensed under the Creative Commons Attribution- Share Alike 3.0 Unported license; Fig 3.21 Martin Groll / licensed under the Creative Commons Attribution 3.0 Germany license; Fig 3.22 Boschfoto / licensed under the Creative Commons Attribution- Share Alike 3.0 Unported license; Fig 3.27 Nigel Chadwick / licensed for reuse under the Creative Commons Attribution- ShareAlike 2.0 license; Fig 3.28 Eric Jones / licensed for reuse under the Creative Commons Attribution-ShareAlike 2.0 license; Fig 3.35 Cmcqueen / licensed under the Creative Commons Attribution 3.0 Unported license; Fig 3.38 Derek Harper / licensed for reuse under the Creative Commons Attribution-ShareAlike 2.0 license; Fig 3.44 © LFM Visuals / Shutterstock.com; Fig 4.13 licensed under the Creative Commons Attribution-ShareAlike 2.5 License; Fig 5.1 Stephen Finn / Shutterstock.com; Fig 5.4 © Soe Than WIN/ AFP/Getty Images; Fig 5.8 Michal Szymanski / Shutterstock.com; Fig 5.12 Fabio Lamanna / Shutterstock.com; Fig 5.13 thomas koch / Shutterstock.com; Fig 5.16 © A Majeed/ AFP/Getty Images; Fig 5.18 © Aref Karimi/AFP/Getty Images; Fig 5.19 Martin Good / Shutterstock.com; Fig 5.22 © Thingrass / Shutterstock.com; Fig 5.24 © Graham Barclay/Bloomberg via Getty Images; Fig 5.25 fpolat69 / Shutterstock.com; Fig 6.4 © ABDELHAK SENNA/AFP/ GettyImages; Fig 6.7 licensed for reuse under the Creative Commons Attribution- ShareAlike 2.0 license; Fig 6.10 Hoang Cong Thanh / Shutterstock. com; Fig 6.11 © DeAgostini/Getty Images; Fig 6.15 licensed for reuse under the Creative Commons Attribution-ShareAlike 2.0 license; Fig 6.16 © LEO RAMIREZ/AFP/GettyImages; Fig 6.19 © Chris Jackson/Getty Images; Fig 6.21 Joseph Sohm / Shutterstock.com;

Fig 6.28 Nilfanion / licensed under the Creative Commons Attribution-Share Alike 3.0 Unported license; Fig 6.29 © Jeff J Mitchell/Getty Images; Fig 6.30 © Aerial Photography Solutions; Fig 6.31 © Jeff J Mitchell/Getty Images; Fig 6.33 Nk.sheridan / licensed under the Creative Commons Attribution-Share Alike 3.0 Unported license; Fig 6.34 Mick Garratt / licensed under the Creative Commons Attribution-Share Alike 2.0 Generic license; Fig 6.36 © Geography Photos/ Universal Images Group via Getty Images; Fig 6.37 Lewis Clarke / licensed for reuse under the Creative Commons Attribution-ShareAlike 2.0 license; Fig 6.40 N Chadwick / licensed for reuse under this Creative Commons License; Fig 7.2 © Humphrey Spender/Picture Post/Getty Images; Fig 7.3 © Monty Fresco/Topical Press Agency/Getty Images; Fig 7.4 © Monty Fresco / Stringer / Getty Images; Fig 7.5 © Jeff J Mitchell/Getty Images; Fig 7.6 Chris Upson / licensed for reuse under the Creative Commons Attribution-ShareAlike 2.0 license; Fig 7.7a Stephen Sweeney / licensed for reuse under this Creative Commons License; Fig 7.7b Leslie-Barrie / licensed for reuse under this Creative Commons License; Fig 7.8 C-L-T-Smith and Gordon-Dowie / licensed for reuse under this Creative Commons License; Fig 7.9 © Bob Thomas/ Getty Images; Fig 7.10 © Jeff J Mitchell/Getty Images; Fig 7.11 Thomas Nugent / licensed for reuse under this Creative Commons License; Fig 7.12 licensed for reuse under this Creative Commons License; Fig 7.14 © London Stereoscopic Company/Getty Images; Fig 7.16 licensed for reuse under this Creative Commons License; Fig 7.17 Cornfield / Shutterstock.com; Fig 7.18 edward-mcmaihin / licensed for reuse under this Creative Commons License; Fig 7.22 licensed for reuse under this Creative Commons License; Fig 7.24 licensed for reuse under this Creative Commons License; Fig 7.28 licensed for reuse under this Creative Commons License; Fig 7.30 © dpa picture alliance / Alamy; Fig 7.31 Celso Pupo / Shutterstock.com; Fig 7.32 T photography / Shutterstock.com; Fig 7.33 licensed for reuse under this Creative Commons License; Fig 7.34 licensed for reuse under this Creative Commons License; Fig 7.35 © Mario Tama/Getty Images; Fig 8.2 Shannon/ licensed under the Creative Commons Attribution-Share Alike 4.0 International, 3.0 Unported, 2.5 Generic, 2.0 Generic and 1.0 Generic license; Fig 8.7 © The Denver Post via Getty Images; Fig 8.12 Adbar / licensed under the Creative Commons Attribution-Share Alike 3.0 Unported license; Fig 8.13 Adbar / licensed under the Creative Commons Attribution- Share Alike 3.0 Unported license; Fig 8.15 Adam Kliczek, http:// zatrzymujeczas.pl (CC-BY-SA-3.0); Fig 8.30 © DigitalGlobe via Getty Images; Fig 8.33 Alan D. Wilson / licensed under the Creative Commons Attribution-Share Alike 2.5 Generic license; Fig 8.34 CityPeak / licensed under the Creative Commons Attribution-Share Alike 3.0 Unported license; Fig 8.35 Steve / licensed under the Creative Commons Attribution-Share Alike 2.0 Generic license; Fig 8.38 Jim Richardson/National Geographic/Getty Images; Fig 8.40 © Annie Griffiths Belt/National Geographic/Getty Images; Fig 9.13 ChinaFotoPress/Getty Images; Fig 9.14 © NOORULLAH SHIRZADA/AFP; Fig 9.15 © Naashon Zalk/Getty Images News; Fig 9.16 © Thomas Imo/Photothek / Getty Images; Fig 10.5 © Roland Neveu / Getty Images; Fig 10.9 © SAID KHATIB/AFP/Getty Images; Fig 10.20 © BRENDAN SMIALOWSKI/AFP/Getty Images.

All other images © Shutterstock.com or public domain.

Text acknowledgements provided on page 254

This product uses map data licensed from Ordnance Survey © Crown copyright and database rights (2014) Ordnance Survey (100018598) – Fig 11.1 and 11.7

The following exam-style questions were adapted from SQA questions with permission, Copyright © Scottish Qualifications Authority: Atmosphere questions 2 and 3; Hydrosphere questions 1 and 2; Lithosphere questions 1 and 2; Biosphere question 1; Population questions 1 and 2; Rural questions 1 and 2; Urban question 2; River basin management questions 1 and 2; Development and health questions 1, 2, 3 and 4.

Whilst every effort has been made to trace the copyright holders, in cases where this has been unsuccessful, or if any have inadvertently been overlooked, the Publishers would gladly receive any information enabling them to rectify any error or omission at the first opportunity.

Contents

Physical Environments

1 Atmosphere

Within the context of the Atmosphere you should know and understand:

- Global heat budget
- Redistribution of energy by atmospheric and oceanic circulation
- Causes, characteristics and impact of the Inter Tropical Convergence Zone (ITCZ).

You also need to develop the following skills:

- Interpret a wide range of numerical and graphical information
- Analyse and synthesise information from a range of numerical and graphical sources.

Make the link

You may have learned more about these elements in Chemistry.

What is the atmosphere?

The atmosphere was originally formed as the Earth cooled; it is a mixture of oxygen (21%), nitrogen (78%), carbon dioxide (0.037%) and other gases such as hydrogen, helium, argon, neon, krypton, xenon and ozone. It also contains water vapour.

These gases are the most dense at the Earth's surface and get less dense as altitude increases. The first 14.5 km (9 miles) from the Earth's surface up contains 90% of the atmosphere's weight, meaning that at the very top of the atmosphere there is only a very thin layer protecting all of the life on the planet.

The global heat budget

The amount of heat that the Earth absorbs and the amount of heat that escapes from the Earth (as radiation) is exactly the same. We can tell this because if the Earth absorbed more than it lost then the planet would be getting warmer with each year, and if more escaped than was absorbed it would be getting cooler. This balance between the incoming and outgoing heat is called the global heat budget (Figure 1.2).

85 km

50 km

Thermosphere

20 km

Mesosphere

Stratosphere

Troposphere

Figure 1.1: *The atmosphere*

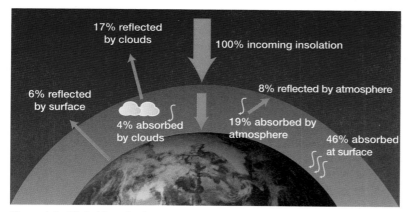

Figure 1.2: *Global heat budget*

- Incoming insolation = 100%
- 46% absorbed at surface
- 6% reflected by surface
- 19% absorbed by atmosphere
- 8% reflected by atmosphere
- 4% absorbed by clouds
- 17% reflected by clouds.

From the diagram we can see that less than half of the insolation actually reaches the Earth – only 46%. Here, the energy is converted back into heat energy and it warms the ground. As the ground warms up, it radiates energy back into the atmosphere. This outgoing radiation energy is called **terrestrial radiation** and is longwave or infrared.

The main reasons for up to 54% insolation being lost are:

Scattering: This occurs when light passes through a transparent medium, which contains small obscuring particles – therefore some light is scattered.

Absorption: The atmosphere absorbs a relatively small amount of insolation, but a great deal of terrestrial radiation. Most of this is absorbed by CO_2 and water vapour.

Reflection: The clouds, the atmospheric gases and the Earth's surface can reflect insolation. Most is reflected by the clouds, but the ice and snow sheets on the Earth's surface also reflect much insolation.

Make the link

You may have learned more about the exchange of energy in Physics.

The ratio between insolation and the amount reflected is called the **albedo**. Some examples are shown in the table below.

Surface	Albedo
Thin clouds (cirrus)	30–40%
Thicker stratus clouds	50–70%
Cumulonimbus clouds	90%
Dark soil	10%
Coniferous forest and urban areas	15%
Grasslands and deciduous	40%
Fresh snow	85%

Figure 1.3: *Cirrus clouds are thin and have a relatively low albedo*

Figure 1.4: *Cumulonimbus clouds are thick and have a high albedo*

Why does insolation vary across the globe?

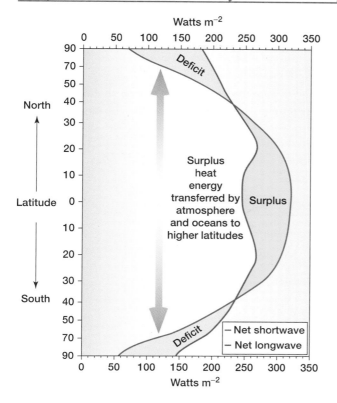

Figure 1.5: *Insolation across the globe*

Figure 1.6: *The curvature of the Earth affects the insolation received in different places*

As shown in Figure 1.5, there is more incoming radiation (net shortwave) than there is outgoing (net longwave) at the Equator, resulting in a surplus of energy. This surplus of energy is transferred to the poles where there is more outgoing radiation than there is incoming, resulting in a deficit of energy.

There are several reasons why insolation varies across the globe.

> ## 🔍 Hint
>
> Even if an exam question does not tell you to draw a diagram, it can still be helpful to include one. For a question on the variance of insolation, a diagram like Figures 1.7 and 1.8 could help you explain the factors.

1. Curvature of the Earth

Figure 1.7 shows that the Earth is a sphere and because of this there are variations in the amounts of insolation received in different places. Both bands of insolation are the same width and therefore the same strength, yet the band that is near the **North Pole** has to heat a **larger area** while the band at the **Equator** can concentrate its heat in a **smaller area**.

This means the insolation at the **Arctic Circle** is stretched and therefore weaker and, conversely, the insolation at the Equator is more concentrated and is subsequently warmer.

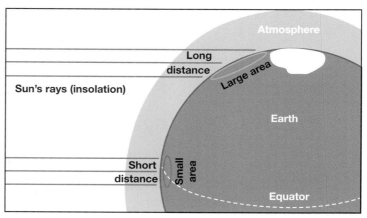

Figure 1.7: *The effect of latitude on insolation*

The same amount of sunlight spreads over a larger area near the poles than near the Equator. When spread over a larger area, the insolation per unit area is decreased, leading to cooler temperatures away from the Equator.

2. Thickness of the atmosphere

The farther from the Equator, the greater the amount of atmosphere the insolation has to penetrate. Therefore, more is lost through the action of **scattering, absorption and reflection**. The thickness of atmosphere is comparatively thinner at the Equator than at the Arctic Circle. This is again due to the Earth being spherical (see Figure 1.8).

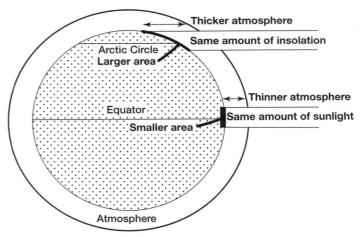

Figure 1.8: *Thickness of the atmosphere*

3. Albedo

The surface of the Earth also affects the heat balance. The **poles** have a **high albedo**, due to the ice caps being light/white in colour, which results in heat loss through terrestrial radiation. They reflect heat. The **equatorial areas** have a **low albedo** due to the dark green and brown of the forests. They absorb heat.

Figure 1.9: *Equatorial areas have a low albedo*

Figure 1.10: *The poles have a high albedo*

4. Tilt of the Earth – seasonal

As the Earth tilts on its axis, the angle of the Sun's rays changes throughout the year. This has an effect on the amount of insolation received by different areas and at different times of the year.

At the winter solstice on 21 December, the North Pole is tilted away from the sun and the overhead sun is found at the Tropic of Capricorn. This means that the North Pole receives no insolation at this time and the Tropic of Capricorn its maximum amount. By 21 March, the Earth has moved into a position where the maximum overhead sun is over the Equator, which receives the most insolation. The Northern and Southern Hemispheres are in spring and autumn respectively. By 21 June, as the Earth continues to move, the maximum overhead sun is now over the Tropic of Cancer. This means the Northern Hemisphere is receiving maximum insolation. Due to the Earth's curve, the South Pole is receiving no insolation at this time.

Figure 1.11: *The Earth tilts on its axis*

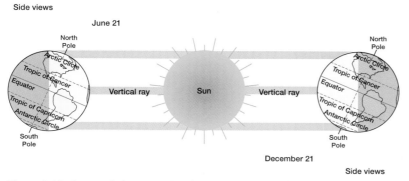

Figure 1.12: *In summer, it never gets dark at the North Pole*

> ## 🔍 Hint
>
> Many questions at Higher need explanation. To gain marks your answer needs reasons.

The tropical areas always receive insolation, no matter what time of year.

Figure 1.13: *Seasonal changes in insolation*

Atmospheric circulation

Make the link

You will remember learning about how pressure affects weather as part of National 5 Geography.

You have seen that there is surplus energy at the Equator and a deficiency at the poles. As the poles do not get progressively warmer or colder, a redistribution of world heat energy must occur. To create a balance, there is constant energy transfer between the Equator and the poles. On Earth, the atmosphere and the oceans act to transfer energy from one area to another. This, however, is not a straightforward journey from the Equator to the poles. This is because air moves from high to low pressure. The Equator is an area of low pressure and the polar areas are high pressure. This is why the air moves in cells.

Figure 1.14 shows a simple three-cell model of atmospheric circulation. This is also called Ferrel's model.

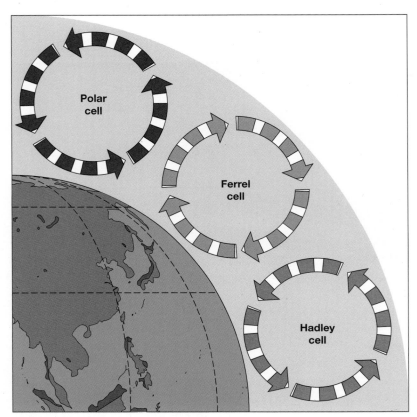

Figure 1.14: *Ferrel's model*

Figure 1.15 shows how this energy is transferred.

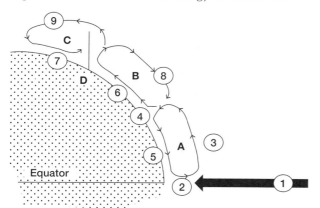

Figure 1.15: *How energy is transferred*

Energy is transferred in the following way:

1. Energy from the Sun heats the air at the Equator.

2. The air rises, which causes low pressure at the Equator.

3. Because of the movement of the Earth, it is pushed towards the poles.

4. Because of the higher latitudes, the air cools and sinks to Earth. The sinking air causes high pressure at the tropics.

At this point, the air can be drawn in one of two directions towards a low pressure zone, either southwards to the equatorial low (5) creating the **trade winds**, or northwards to the polar front creating the **westerlies** (6).

5. Air can be drawn back to the equatorial low pressure. This completes the Hadley cell.

6. Some air moves polewards.

7. Here it meets polar air and forms the polar front at **D**. This completes the Ferrel cell.

The air is lighter and forced to rise, causing low pressure.

8. Some is diverted to the high pressure zone at (4).

9. And some to the polar high pressure, creating the **polar easterlies**.

Figure 1.16: *The earliest wind charts of the world's oceans were created in the mid-1800s and were used by traders to plot their journeys*

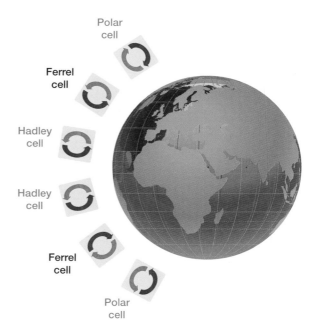

Figure 1.17: *Ferrel, Hadley and Polar cells*

The three cells together

The Hadley cell (**A** in Figure 1.15) exists because of the solar energy heating the air at the Equator and causing the equatorial low pressure, which draws in air from the sub-tropical high pressure. The Ferrel cell (**B**) feeds warm air to the low latitudes and draws cold air back. The Polar cell (**C**) is caused by cold, dense air sinking at the poles and forming an area of high pressure. Where there is an area of convergence where cells meet, there are high winds.

Why do the cells occur where they do?

- Wind always moves from **high** to **low** pressure.
- There are two main areas of high pressure in the world – the tropics and the polar highs.
- There are two main areas of low pressure in the world – the equatorial low and the polar front.
- Air moves between these four areas.
- Low pressure occurs where the air is forced to rise upwards.
- At the Equator, the air rises as it is heated by the Sun.
- Near the poles and the zone of convergence, the air rises because the warmer air from the lower latitudes is pushed upwards.
- High pressure occurs where air falls. This is because it has cooled and has moved away from the Equator.

World pressure and wind patterns – the Coriolis force

The movement of air produces wind. These wind patterns show how the air moves. It always moves from high to low pressure. As the Earth rotates, other factors kick into force. One of these is the Coriolis force. As the Earth spins, we experience a force known as the Coriolis force. This deflects the direction of the wind to the right in the Northern Hemisphere and to the left in the Southern Hemisphere. This is why the wind-flow around low- and high-pressure systems circulates in opposing directions in each hemisphere (see Figure 1.18).

> ### ✦: Make the link
>
> If you study Physics you will learn more about the actions of forces.

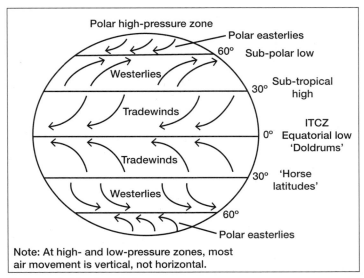

Figure 1.18: *The Coriolis force*

The Earth does not sit in line with the Sun all year – as we have already seen, it tilts. This tilting movement also affects how the air circulates around the globe. In June/July, the Northern Hemisphere tilts towards the Sun so receives most of the Sun's energy, while in December the Southern Hemisphere receives most of the energy. This means that the equatorial low pressure moves up and down, keeping in line with the maximum insolation – the **thermal equator** (Figure 1.13).

Oceanic circulation

Ocean waters are a vast heat source for the atmosphere. The surface waters of the oceans are deeply penetrated by insolation, providing warming to the depths of up to 100 m. Oceans cover 70% of the Earth's surface, and therefore receive over two-thirds of global insolation. As with the atmosphere, circulation systems act to transfer energy from the tropics to the poles. Surface ocean currents are driven largely by the major wind systems. Like the cells in the atmosphere, ocean water circulates in gyres or loops, controlled by the wind systems. The waters near the Equator receive more heat than those near the poles. Therefore, warm water flows outwards from the equatorial regions towards higher latitudes. In consequence, the colder water from the poles flows towards warmer regions, creating a circulatory system. In a model Earth with no land masses, the ocean currents would look like Figure 1.20.

Figure 1.19: *Oceans receive over two-thirds of global insolation*

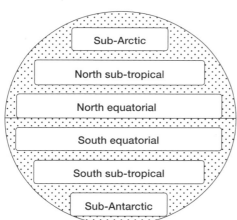

Figure 1.20: *The position of the ocean currents if there were no land masses*

The most important points to remember are:

- Warm water is transferred from the tropics to the poles at the surface.
- Oceans transfer cold water from the poles to the Equator.
- Ocean currents broadly follow wind patterns.
- The currents flow clockwise in the north and anti-clockwise in the Southern Hemisphere.

In practice, however, land masses disrupt this pattern, giving a much more complicated pattern (Figure 1.21).

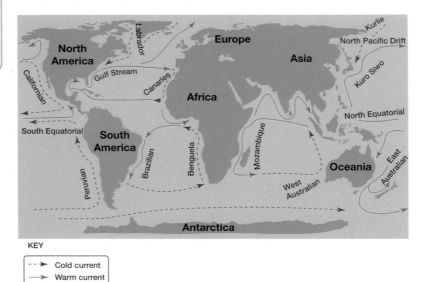

Figure 1.21: *The ocean currents*

Effects of ocean currents

Ocean currents can have a very significant effect on climate. For example, the UK has relatively mild, wet winters due to the effect of the North Atlantic Drift, which is an extension of the Gulf Stream. Winds blowing over this current are warmed by it, so can absorb more moisture.

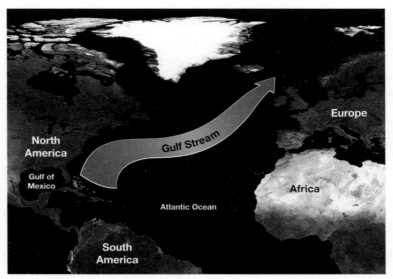

Figure 1.22: *The Gulf Stream*

In contrast, the trade winds blowing over the cold currents of the eastern Atlantic (Canaries Current) are cooled so become stable and dry. This provides the arid environments of coastal Northwest Africa around the Sahara.

In summary:

- Ocean currents follow loops or gyres.
- Ocean currents move clockwise in the North Atlantic.
- In the Northern Hemisphere, the loop is formed by warm water from the Gulf of Mexico travelling northward (Gulf Stream).
- Colder water moves southward (Canaries Current).
- Currents from the poles to the Equator are cold currents, e.g. Labrador Current.
- Currents from the Equator to the poles are warm currents.
- Ocean currents are greatly influenced by winds and land masses, which deflect the ocean currents.
- The movement of warm and cold water helps to redistribute energy.

El Niño

A change in the pattern of ocean currents can have a devastating effect on the fishing industry. El Niño is a periodic change in the currents. This means that the Equatorial Current, which moves west across the Pacific, fails to do so. The winds die and a large pool of warm water in the Pacific moves east. This warm water moving towards Peru prevents the cool water up-welling and as a result many fish and seabirds die and the fishing industry suffers badly.

🔍 **Hint**

Marks can be gained for describing different ocean currents.

Figure 1.23: *Due to the Gulf Stream, the UK has wet and mild winters*

Figure 1.24: *The trade winds create dry and hot conditions over the Sahara*

Figure 1.25: *Fishermen in the Philippines protesting that the government has not done enough to help them following a prolonged dry spell created by El Niño*

Air masses and ITCZ

Global wind circulation and ocean currents are important in determining climate patterns. These are the origins of the air masses that affect our weather. In some areas of the world, air accumulates and adopts uniform characteristics of temperature and humidity. Their characteristics remain similar as the air moves away from the source. Over Africa, the north-east trades bring the tropical continental air mass and the south-east trades bring the tropical maritime air mass.

A tropical maritime air mass originates in tropical latitudes over the Gulf of Guinea in the Atlantic Ocean. It brings warm, moist, unstable air with high levels of humidity. It is moist, as it has travelled across the ocean, picking up moisture as it moves. Because it originates in tropical latitudes, it brings hot conditions.

A tropical continental air mass originates over the land, e.g. the Sahara Desert, in tropical latitudes. It brings warm, dry, stable conditions with low levels of humidity as it travels over land as opposed to water. In both summer and winter it brings hot and dry conditions.

Where these two air masses meet, they form the Inter Tropical Convergence Zone (ITCZ).

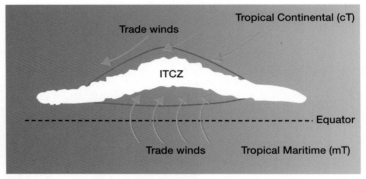

Figure 1.26: *Bird's-eye perspective of the Inter Tropical Convergence Zone (ITCZ)*

ITCZ around Africa

Tropical Continental air from the Sahara prevents the Tropical Maritime air mass from rising – thus it cannot rise, cool and condense to form rain. Rainfall only occurs where the compressed edge (also known as the squall line) reaches around 200 m. This usually occurs 400 km south of the actual position of the ITCZ. The weather associated with the ITCZ is towering cumulonimbus clouds, heavy precipitation and high intensity rainfall (150–200 mm per hour).

Movement of the ITCZ

We know that the zone of maximum insolation (thermal equator) is not fixed. It moves north in June and south in January because of the tilt of the Earth. It follows then that the ITCZ also moves north in summer and south in winter (Figure 1.27).

Make the link

In the Rural and River basin management chapters you will learn about how a lack of rainfall can affect a region, and how water can be managed.

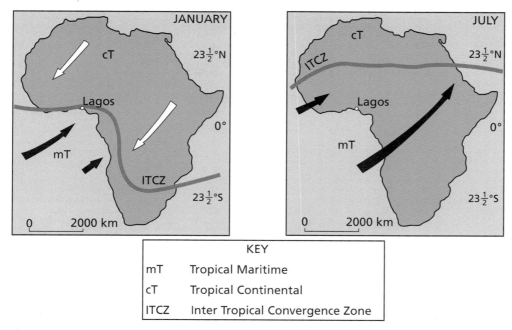

Figure 1.27: *Movement of the ITCZ*

ITCZ, climate and West Africa

The climate is highly **seasonal** in West Africa due to the movement of the ITCZ.

In the winter, the dry season, the dominant air mass is tropical continental, cT, with the dry Harmattan wind blowing across nearly all of West Africa. It lowers the humidity, which can lead to hot days and cool nights. Only the coastal south remains under the influence of the mT air. In June/July the thermal equator moves north to the Tropic of Cancer, dragging with it the ITCZ. This brings with it the convectional rainfall of the ITCZ. Following on behind is the mT air mass, which is carrying large amounts of potential precipitation. This pattern results in a marked decrease in rainfall away from the Equator, with the rainfall occurring at different times of the year.

> ### 🔍 Hint
>
> Directly underneath the ITCZ = convectional thunderstorms.
>
> South of the ITCZ = wet weather from mT.
>
> North of the ITCZ = dry weather from cT.

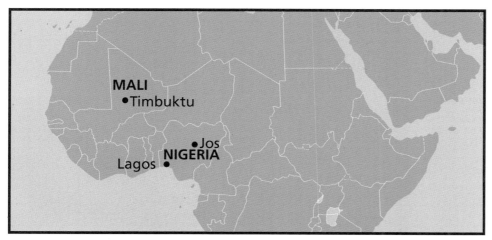

Figure 1.28: *Location of Lagos, Jos and Timbuktu*

	J	F	M	A	M	J	J	A	S	O	N	D
Lagos	36	41	132	163	290	452	294	53	154	200	66	28
Jos	3	4	19	98	182	200	302	296	222	46	4	1
Timbuktu	0	0	3	0	3	25	80	82	50	3	0	0

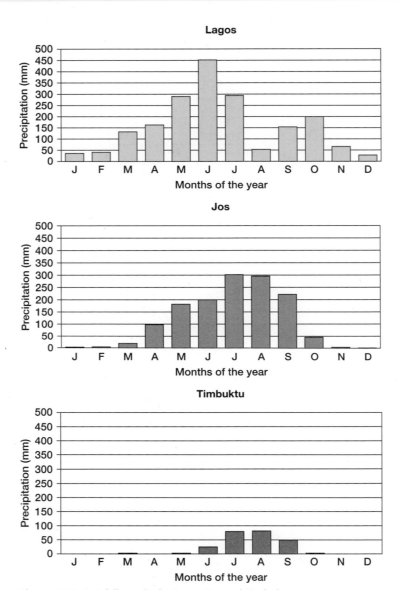

Figure 1.29: *Rainfall graphs for Lagos, Jos and Timbuktu*

Rainfall in West Africa

Rainfall varies dramatically over the region and can be described using the three 'i's: **incidence, intensity** and **irregularity**.

1. Incidence is the variability within the year.

Seasonality is the key word here. Areas have a distinct wet and dry season. The wet season falls later in the year the further north you go.

Areas in the south have a twin maxima (double peak) caused by the ITCZ passing over twice. The twin precipitation peaks occur because the ITCZ moves north in the early part of the year, bringing rainfall, before moving south later in the year, again bringing rainfall.

The southern areas receive a far higher total than those in the north. This is because even when the ITCZ has moved north, the mT air mass is still overhead – bringing mild, wet weather to the southern areas.

2. Intensity means how heavy the rainfall is.

One of the problems of high intensity is that much of the water will be lost to overland flow and run-off.

3. Irregularity refers to the extent that rainfall varies from year to year compared to the long-term average.

In much of West Africa, variability is similar to the UK (16–20% from the mean), but it can vary by up to 60%. The graph in Figure 1.30 shows how variable rainfall can be.

>
> ### Make the link
>
> In the Hydrosphere chapter you will learn more about rainfall and the hydrological cycle.

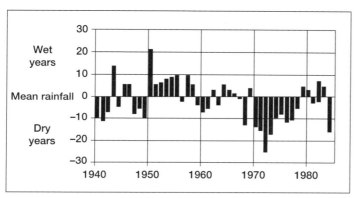

Figure 1.30: *Rainfall in the Sahel*

This has had a profound effect on the environment. During the 1950s the mT air mass was powerful, keeping the ITCZ in higher latitudes. This increased rainfall encouraged settlement north to the margins of the Sahara (Sahel region). However, since then there has been a southward shift of the ITCZ some 200–300 km, possibly due to global warming.

> ### Make the link
>
> You will look at the causes of global warming in detail in the Global climate change chapter.

This means that land that was previously able to support agriculture can no longer do so. In 1974, one million people migrated southwards in West Africa, increasing population density there. The strength of the cT air mass has resulted in an extended dry season, and this has had major effects on agriculture, with people being forced to adapt to new conditions and grow crops with which they may not be familiar.

Figure 1.31: *There were severe droughts in West Africa in 2010 and 2012*

In summary:

- The coastal and southern areas of Africa lie south of the ITCZ for most of the year.
- This area receives moist mT air for most of the year.
- mT air originates over the Atlantic Ocean and brings hot, wet weather.
- This area can have twin rainfall peaks as the ITCZ moves north and south.
- The northern areas of Africa lie north of the ITCZ in winter.
- This area receives dry cT air.
- In summer, the ITCZ moves north and brings moist (mT) air – giving rain.
- Rainfall becomes more and more limited the farther north you go.

Summary

In this chapter you have learned:

- The makeup of the Earth's atmosphere
- The components of the heat budget
- The causes for the loss of insolation from the Earth
- The reasons for variations in insolation across the surface of the Earth
- How heat is redistributed over the globe
- The role of Ferrel, Polar and Hadley cells in the redistribution of heat over the globe
- Wind patterns over the globe
- Ocean currents and their effects
- The properties of tropical continental and tropical maritime air masses and the formation of the ITCZ
- How the ITCZ affects the climate of West Africa
- The effects of the ITCZ on West Africa.

You should have developed skills and be able to:

- Interpret tables and diagrams
- Analyse the movement of air masses using cell diagrams
- Identify ocean and wind patterns from a map
- Use maps to explain climate patterns
- Interpret climate graphs.

End of chapter questions and activities

Quick questions

1. Explain why up to 46% of the Sun's heat is lost to the Earth.

2. Explain, in detail, the variation in the amount of solar energy received between the tropics and the poles.

3. Explain why air moves between high and low pressure.

4. Describe and explain a Ferrel, Polar and Hadley cell.

5. Explain how the three cells together transfer energy around the world.

6. Explain, in detail, how the pattern of ocean currents in the North Atlantic Ocean helps to maintain the global energy balance.

7. Discuss the origin, nature and characteristics of the maritime tropical and continental tropical air masses.

8. Account for the variation in rainfall within West Africa.

> **🔍 Hint**
> Put as much detail into an answer as possible. Detail improves grades.

> **🔍 Hint**
> In some questions, description is necessary but most marks come from explanation.

Exam-style questions

1. With the aid of an annotated diagram or diagrams, **explain** why there is a surplus of solar energy in the tropical latitudes and a deficit of solar energy towards the poles.

8 marks

2. Study Figure 1.32.

   Explain why the Earth's surface absorbs only 50% of the solar energy received at the edge of the atmosphere. You should refer to both conditions in the Earth's atmosphere and at the Earth's surface.

6 marks

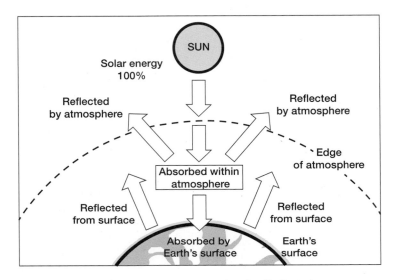

Proportion of solar energy absorbed/reflected

RS = reflected from surface

Figure 1.32: *Earth/atmosphere energy exchange*

3. Study Figures 1.33 and 1.34.

   a. Describe the variation in rainfall within West Africa.

   b. Explain the reasons for the variation.

8 marks

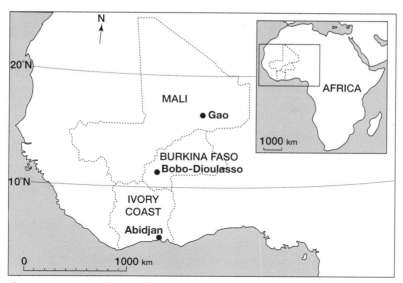

Figure 1.33: *Map of West Africa*

Bobo-Dioulasso: total precipitation—1000 mm

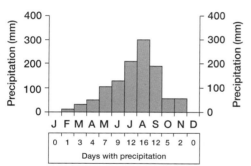

Abidjan: total precipitation—1700 mm

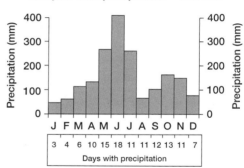

Gao: total precipitation—200 mm

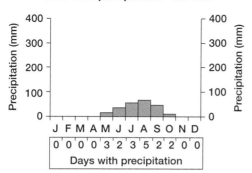

Figure 1.34: *Average monthly rainfall/days with precipitation*

Activity 1: Oral presentation

In groups, create a presentation on the effects of the ITCZ in West Africa. Your presentation should be roughly 10 minutes in length and cover effects on both people and the environment. You can use PowerPoint to help if you wish. Be prepared to answer questions from other pupils at the end of your presentation.

Activity 2: Revision cards

In pairs, collect small pieces of card and make up revision cards. Use the glossary for this chapter and write each word on the front of the card and its definition on the back. Test each other to see if you can explain each word.

Learning Checklist

You have now completed the Atmosphere chapter. Complete a self-evaluation sheet to assess what you have understood. Use traffic lights to help you make up a revision plan to help you improve in the areas you have identified as amber or red.

- Demonstrate an understanding of the Earth's atmosphere.

- Explain the balance between incoming and outgoing heat in the Earth's global heat budget.

- Discuss the reasons why 46% of the heat is lost to the Earth, mentioning the terms scattering, absorption and reflection.

- Explain the reasons why there is a surplus of energy at the Equator and a deficit at the poles. You should refer to the curvature of the Earth, thickness of the atmosphere, albedo and the tilt of the Earth.

- Explain why air moves between high and low pressure.

- Describe and explain a Ferrel, Polar and Hadley cell.

- Discuss how the three cells together transfer energy around the world.

- Describe and explain global wind patterns.

- Describe the distribution of ocean currents.

- Explain how the oceans transfer energy.

- Explain the effects of ocean currents.

- Understand the characteristics of a tropical maritime and tropical continental air mass and the effects of these two air masses meeting.

- Explain why the ITCZ moves.

- Using maps and diagrams and specific examples, explain how the movement of the ITCZ affects the climate of West Africa.

- Discuss the effects of this movement of the ITCZ on the people and landscape of West Africa.

Glossary

Absorption: the process by which radiant energy is retained by the atmosphere.

Air mass: a body of air with uniform weather conditions, such as similar cloud types, temperature and humidity.

Albedo: a measure of how much light that hits a surface is reflected without being absorbed.

Atmosphere: a blanket of gases that contains solid material, such as volcanic dust and blown soils, and is attached to the Earth by the force of gravity.

Climate: average weather conditions over at least 35 years.

Coriolis force: the Earth's rotation creates an energy force that deflects the direction of the wind to the right in the Northern Hemisphere and to the left in the Southern Hemisphere.

Cumulonimbus clouds: a dense towering vertical cloud associated with thunderstorms.

Doldrums: a belt of calms and light winds between the northern and southern trade winds of the Atlantic and Pacific oceans.

El Niño: an abnormal warming of surface ocean waters in the eastern tropical Pacific.

Front: the boundary between two air masses.

Global heat budget: the balance between incoming and outgoing heat.

Harmattan: hot, dry wind that blows from the north-east or east in the Western Sahara and is strongest in late autumn and winter.

Incidence: the variability within the year.

Insolation: incoming heat from the Sun.

Intensity: how heavy the rainfall is.

Irregularity: the extent that rainfall varies from year to year compared to the long-term average.

ITCZ: Inter Tropical Convergence Zone – the region where the north-easterly and south-easterly trade winds converge, forming a continuous band of clouds or thunderstorms near the Equator.

Loops: large system of rotating ocean currents.

Migration: seasonal movement of the ITCZ.

Precipitation: any form of moisture but most commonly rainfall.

Prevailing winds: the most frequent wind direction a location experiences.

Sahel: semi-arid area south of the Sahara.

Scattering: light passing through a transparent surface, which contains small obscuring particles – therefore some light is scattered.

Seasonal: periodic.

Thermal equator: zone of maximum insolation.

Trade winds: any of the nearly constant easterly winds that dominate most of the tropics and sub-tropics throughout the world, blowing mainly from the north-east in the Northern Hemisphere, and from the south-east in the Southern Hemisphere.

Tropical continental: warm dry air mass.

Tropical maritime: warm wet air mass.

Warm current: body of warm water.

2 Hydrosphere

Within the context of the Hydrosphere you should know and understand:

- The formation of erosional and depositional features in river landscapes
- The hydrological cycle within a drainage basin
- The interpretation of storm hydrographs.

You also need to develop the following skills:

- Interpret numerical and statistical data
- Analyse information from graphs
- Interpret information from maps.

The effects of rivers on the landscape

Most rivers have their sources up in the mountains. A river has three stages or courses: the upper course, the middle course and the lower course.

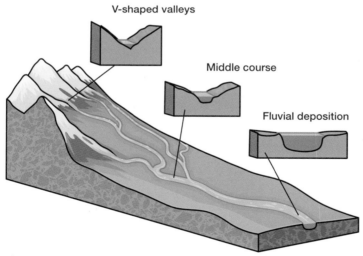

Figure 2.1: *A long profile of a river valley*

A river uses three main processes to alter the landscapes it flows through: erosion, transportation and deposition.

Erosion

Erosion is when the river wears away at its bed and banks. The amount of erosion depends on the river's energy. This is dependent on the amount of water in the river, its speed and seasonal variation.

Erosion in a river is the result of four processes. These are:

1. **Hydraulic action** – the sheer force of the river causes particles to break off from the riverbanks and bed. This happens as air becomes trapped in the cracks of the riverbank and causes the rock to break off.

2. **Attrition** – when boulders, rocks and pebbles crash into each other and bits break off, making them more rounded and smaller in size.

3. **Corrasion/Abrasion** – the wearing away of the riverbanks and bed by the river's load.

4. **Corrosion** (chemical solution) – when the river water dissolves minerals from the rocks and washes them away.

The river's valley is deepened by vertical erosion (downwards) and widened by lateral erosion (sideways).

Transportation

This is where material is moved from one place in the river and deposited in another part of the river.

A river transports its load in four ways:

1. **Traction** – when large boulders and rocks are rolled along the riverbed.

2. **Saltation** – when small pebbles and stones are bounced along the riverbed.

3. **Suspension** – when fine, light material such as alluvium is carried along in the water.

4. **Solution** – when minerals are dissolved in the water and carried along in solution.

Deposition

Deposition occurs when a river loses energy and can no longer carry its load. This can occur when there is a decrease in gradient, an increase in its load or the channel is widening or meandering.

River landscape features

A river can be divided into three courses: the upper, middle and lower courses. In each course, the river forms different river features.

Upper course: V-shaped valley

Stage 1: Valleys in the upper course are narrow and steep sided as rivers in this course only have enough energy to erode downwards. This is called vertical erosion and involves processes like abrasion.

Stage 2: Abrasion scrapes bedload against the riverbed, eroding it. As the river cuts down, the valley sides are attacked by freeze-thaw weathering, which breaks up and loosens the soil and rock.

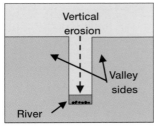

Figure 2.2: *Stage 1*

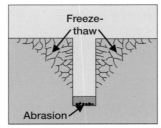

Figure 2.3: *Stage 2*

Stage 3: Gravity causes the weathered material to slip down the slope; it is then washed into the river and carried downstream. This process steepens the valley sides.

Stage 4: The narrow, steep-sided valley that is created is typically 'V' shaped and is common around upland rivers.

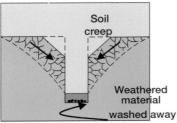

Figure 2.4: *Stage 3*

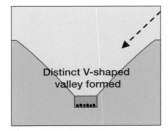

Figure 2.5: *Stage 4*

Figure 2:6: *V-shaped valley*

Upper course: Waterfall

Stage 1: Waterfalls are commonly found in the upper course of a river where differential erosion takes place. Bands of harder, more resistant rock such as limestone are found overlying softer rock such as mudstone.

Stage 2: The power of the water erodes the softer rock faster than the harder rock by hydraulic action. Hydraulic action is a process where air is compressed into the riverbank causing materials to be dislodged. A plunge pool forms as a result. The softer rock is more easily worn away and the hard rock is undercut, leaving an overhang above the plunge pool.

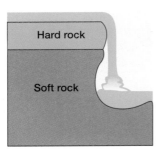

Figure 2.7: *Stage 1*

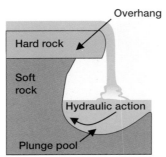

Figure 2.8: *Stage 2*

Stage 3: Over time, the hard rock is left unsupported and collapses due to gravity into the plunge pool. Rock fragments swirling around deepen the plunge pool by abrasion (when the force of the water throws bedload against the banks and riverbed, causing erosion). Attrition also occurs here, where the rocks in the plunge pool hit off each other, eroding further.

Stage 4: This process is repeated and gradually the waterfall retreats upstream, leaving behind a steep-sided gorge.

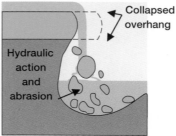

Figure 2.9: *Stage 3*

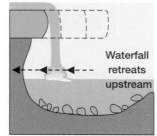

Figure 2.10: *Stage 4*

Figure 2.11: *High Force Waterfall*

Middle/lower course: Meander

Stage 1: Meanders are common in the middle and lower courses of a river. Lateral erosion takes place here as the land is flatter. Pools form in the slower and deeper parts of a river, whereas riffles form in the faster, shallower parts of a river. Where riffles and pools are formed, the water begins to move from side to side in the river channel.

Figure 2.12: *Stage 1*

Stage 2: Erosion takes place at the outside bend of the river, forming a river cliff due to faster flow. Hydraulic action compresses air into the riverbank, causing material to be dislodged and abrasion uses the force of the water to throw bedload against the banks, causing erosion. These processes are assisted by a helicoidal (corkscrew) flow, which removes material and carries it to the opposite bank.

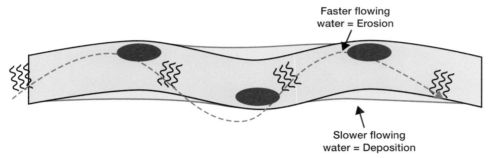

Faster flowing
water = Erosion

Slower flowing
water = Deposition

Figure 2.13: *Stage 2*

Stage 3: Deposition occurs on the inside bends of the river, forming a river beach due to slower flow. Meanders continually migrate downstream due to continuing erosion.

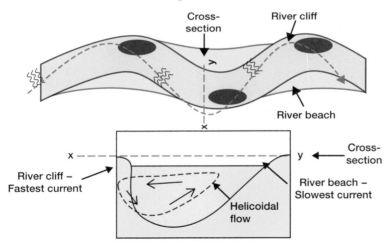

Cross-section

River cliff

River beach

Cross-section

River cliff – Fastest current

River beach – Slowest current

Helicoidal flow

Figure 2.14: *Stage 3*

Lower course: Oxbow lake

Stage 1: In a river's lower course, meanders become more pronounced. The outer banks of a meander continue to erode laterally.

Stage 2: The water flows faster on the outside bend of a meander, so erodes the banks by the processes of hydraulic action and abrasion, causing the banks to become steeper. Deposition takes place on the inside bend of a meander where it has less energy as the water is slow and shallow.

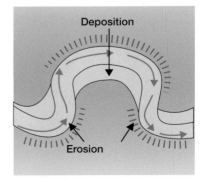

Deposition

Erosion

Figure 2.15: *Stage 1*

Stage 3: The neck of the meander narrows gradually. At a time of flood, the river will cut through this neck, taking the shortest route possible. As the fastest flow is now in the middle of the river, no further erosion takes place.

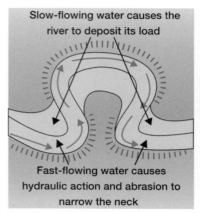

Figure 2.16: *Stage 2*

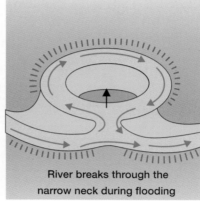

Figure 2.17: *Stage 3*

Stage 4: Deposition gradually seals off the old meander and an oxbow lake is formed. The old meander is left isolated and may eventually dry up.

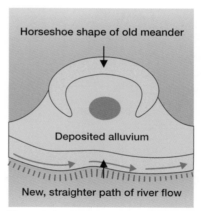

Figure 2.18: *Stage 4*

Figure 2.19: *Meander and oxbow lake*

Why we need water

'Water is the driving force of all nature.'

Leonardo da Vinci

All forms of life on Earth depend on water for survival. However, water is not evenly distributed around our planet. Some areas suffer from droughts while others have frequent floods. In developed countries like the UK and USA, average water consumption per person is high. We use water for drinking, washing, cooking and cleaning. Farming, industries and leisure and recreational activities also use up valuable supplies of water. The average person in the developing world uses only 10 litres per day. Many people in developing countries do not have proper water supplies in their homes.

> **Make the link**
>
> If you take Modern Studies, you may have learned about differences between developed and developing countries.

> **? Did you know?**
>
> The average European uses 200 litres of water every day, while Americans use 400 litres. The average person in the developing world uses only 10 litres per day.

Figure 2.20: *In the developing world, many homes do not have water supplies*

The hydrological cycle

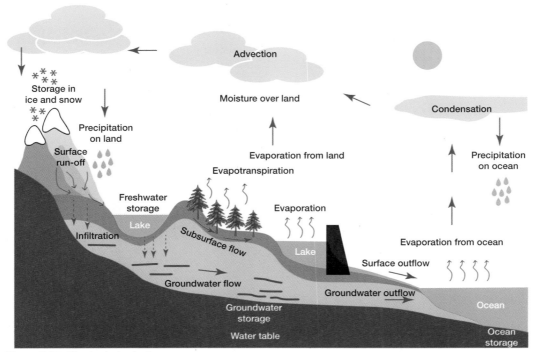

Figure 2.21: *The hydrological system*

🔎 Hint

Explaining the hydrological cycle will involve some description, but remember, most marks come from explanation.

The global hydrological cycle is a major system linking all the elements of our environment by the distribution and movement of water with energy transfers at all points in the system as shown in Figure 2.21.

The Earth contains a set amount of water that cannot be lowered by loss or increased by gain of water. The highest percentage of the Earth's water is found in oceans and seas.

Water is continually moving around the Earth from the sea to the sky to the land and back to the sea again in a continuous cycle. The Sun provides the energy for the cycle. Water is **evaporated** from the sea whereby water as a liquid is converted into water vapour (gas). Water also enters the atmosphere from the process of **transpiration** (water loss from plants) and **sublimation** (water loss from ice). As the water vapour rises, it begins to cool and **condensation** takes place, resulting in the formation of clouds. Clouds travel inland via the process of **advection** and release water as **precipitation**, which returns to the rivers and therefore the sea by through flow. The slowest form of water to return to its source is that of **groundwater**, which sinks deep into the ground before entering the sea. Some precipitation will also be **intercepted** by vegetation and is absorbed in the leaves and roots. Water may also be stored in the water table. The water table is a form of groundwater storage made up of permeable rocks that have had their spaces filled with water. The water table rises and falls depending on the amount of rain. Eventually the water returns to the sea and the cycle begins again.

Figure 2.22: *Water is evaporated from the sea*

Figure 2.23: *Clouds are formed and travel inland*

Figure 2.24: *Some precipitation is intercepted by vegetation*

Figure 2.25: *Eventually the water returns to the sea and the cycle begins again*

River basins

Rivers start in upland areas and flow downhill across the land, ending in lakes or at the sea. Figure 2.26 shows the main features of a river basin. A drainage basin forms part of the hydrological cycle as water is transferred in a continuous cycle from the land to the sea to the atmosphere.

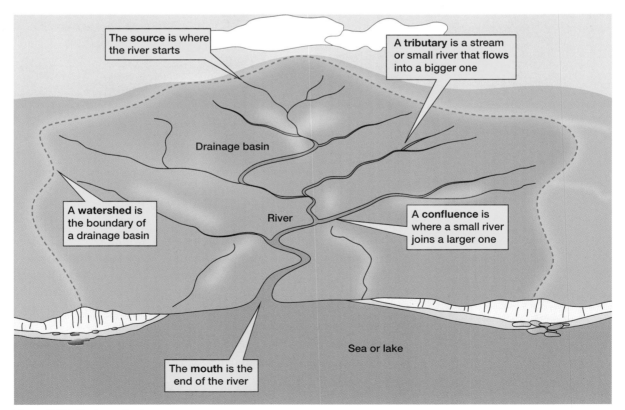

Figure 2.26: *Drainage basin*

A **drainage basin** is the area into which water will flow. Each river system will have its own drainage basin. Drainage basins are separated by **watersheds**. A drainage basin also has inputs, storage, transfers and outputs. These are shown in Figure 2.27.

The main **input** is **precipitation,** mainly in the form of rainfall and snow. The type, the strength, the length and the rate of occurrence all have an impact on the amount of water in the system.

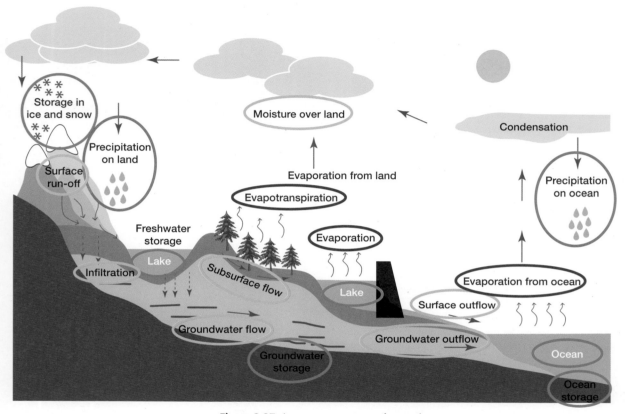

Figure 2.27: *Inputs, storage, transfers and outputs*

Storage is the moisture kept in the system. This can be found in the **soil**, in **lochs**, **ice caps** and **glaciers** and on the **leaves** and **roots** of vegetation. Water seeps into the ground by the processes of infiltration and percolation and is stored in the soil or in rock stores such as the water table.

Transfers are the movements of water through the system, in the air through clouds and precipitation and on the ground through surface run-off, **infiltration** and **groundwater flow**. **Overland flow** transfers water through the basin as sheet wash and across the surface in rivers. **Through flow** is the movement of water through the soil towards the river. **Groundwater flow** is very slow and takes time to flow through the system into the rivers.

An **output** is the moisture that leaves the system by processes such as **evaporation**, **transpiration** and through water flowing into the seas and oceans.

Make the link

If you take Chemistry, you will have learned about processes such as evaporation.

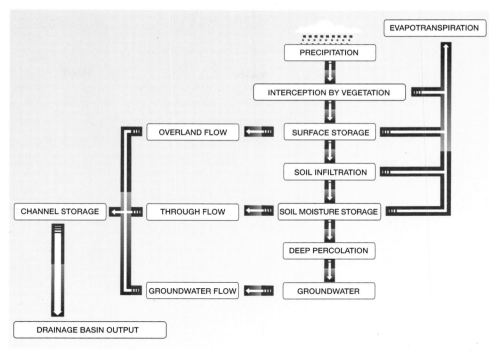

Figure 2.28: *The drainage basin as a system*

Human factors affecting the hydrological cycle

In urban areas there are many areas of impermeable concrete and tarmac surfaces, which reduce infiltration in the soil, and drains that carry rainwater quickly to river channels. This makes flooding more likely. Artificial watering of crops through irrigation removes water from the river, lowering its discharge. Deforestation reduces interception as there are no leaves or tree trunks to soak up the rainwater and less evapotranspiration due to a lack of leaves and vegetation. Deforestation increases discharge and increases flood risk. Afforestation (the planting of trees) has the opposite effect. River management schemes with dams and reservoirs stop the peaks and lows of the river and even out river discharge. Straightening and dredging of channels may speed up channel flow.

Physical factors affecting the hydrological cycle

The amount of precipitation, its intensity, frequency and duration affects how much water enters the cycle. Some soils, such as clays, absorb less water at a slower rate than sandy soils. Soils absorbing less water result in more run-off overland into streams. Water falling on steeply sloped land runs off more quickly and infiltrates less than water falling on flat land.

? Did you know?

Approximately 76% of precipitation falls on the oceans while 24% falls on the land. Of all the water available, 97% is stored in the oceans, 2% in glaciers and ice caps, just under 1% in groundwater, soil water, rivers and lakes and 0·001% in the atmosphere. 16% of water is evaporated from the land, while 84% is evaporated from the oceans.

Make the link

You will learn more about river management schemes in the River basin management chapter and more about deforestation in the Rural chapter.

Hydrographs

What are hydrographs?

The amount of water in a river at any given point is known as its **discharge**. This is measured in cumecs (cubic metres per second). This can be calculated by multiplying the **velocity (speed)** of the river by **channel volume** at a given point along the river at a given time.

Hydrographs are graphs that show **river discharge** over a given period of time and show the response of a drainage basin to a period of rainfall.

Figure 2.29 shows a **hydrograph** and its main features. A hydrograph shows **two graphs** – one for **rainfall** shown by a **bar** graph and one for **discharge** shown by a **line** graph.

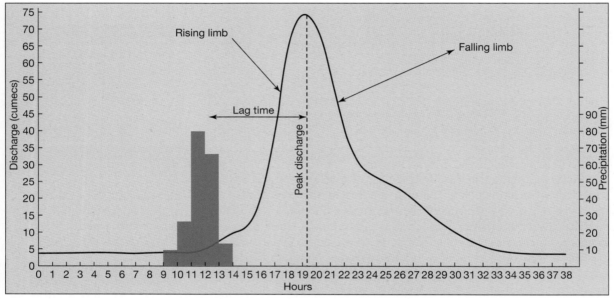

Figure 2.29: *The components of the hydrograph*

A **storm hydrograph** is a specific type of hydrograph that shows precipitation and discharge during and after a storm. Below is a storm hydrograph for the fictional River Corr:

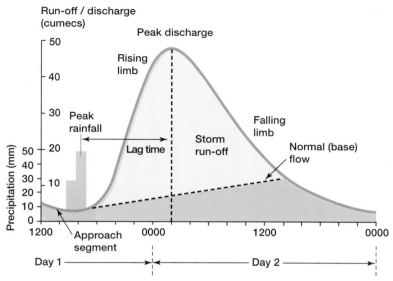

Figure 2.30: *Storm hydrograph*

Hint

If you get a question on hydrographs in the exam, remember to use the appropriate terminology such as 'rising limb', 'peak discharge' and 'lag time'. This will demonstrate your understanding of the graph to the examiner. Make sure you use figures in your answer.

The line shows the discharge of the river and the series of bars shows the precipitation. The rising limb is the climbing part of the discharge line that has an upward trend, indicating that the discharge is increasing. The falling limb is the opposite, showing that the discharge is falling. The lag time is the time difference between the peak precipitation and the peak discharge. A long lag time indicates that it's taking a long time for precipitation to enter the river. Conversely, a short lag time indicates that the precipitation is entering the river fairly quickly.

Make the link

You may have learned how to interpret line and bar graphs in Maths.

Describing the hydrograph

The following points can be mentioned when describing the hydrograph in Figure 2.30.

The river level is fairly low at 1200 hours on Day 1 (before heavy rain) but there is a rapid rise in discharge from 1800 hours due to a rainstorm some five hours previously. Continued heavy rainfall (up to 50 mm) results in more water reaching the river from through flow as well as from surface run-off. The river reaches a peak discharge of 48 cumecs around 0200 hours on Day 2. Rainfall declines sharply and then stops at 1800 hours on Day 1; river levels decline fairly abruptly and then more gradually return to normal (base) flow level by midnight of day 2 at 10 cumecs.

Hint

If you are given a diagram such as a hydrograph in the exam, make sure you study it carefully before writing your answers. Pay attention to the heading, the units used and the key.

Figure 2.31: *A river travelling down a slope*

Make the link

You will study types of soil in detail in the Biosphere chapter.

Make the link

You may remember learning about weather as part of National 5 Geography.

Figure 2.32: *Snowstorms increase a river's discharge, but with a long lag time due to the time it takes the snow to melt*

Factors affecting a storm hydrograph

The drainage basin

The size and shape of the drainage basin affects a hydrograph. Large basins will have high peak discharges because their larger areas will receive more precipitation. However, they will also usually have longer lag times than small basins due to the greater distances the water has to travel to the river.

Basins that contain lots of rivers and streams will have a steep falling limb and a short lag time because this high drainage density allows water to drain out of them more quickly.

Basins with steep slopes will have a high peak discharge and a short lag time because water travels faster down a slope.

Soil and rock type

Rivers that are in areas of impermeable rock will tend to have short lag times and a high peak discharge. The short lag time is because the water is not able to infiltrate the ground and will travel to the river overland, which is quicker than by groundwater flow. The high peak discharge is due to the fact that impermeable rocks do not store water and so all the water will enter the river, rather than some of it being stored underground.

The same is true for the type of soil – impermeable soils such as clay will affect the hydrograph in the same way as impermeable rock. Fine soils will have the opposite effect: a lot of the water will infiltrate the soil and be stored there, resulting in an increased lag time and a lower peak discharge.

Weather and climate

Heavy rainstorms will result in a higher discharge due to the volume of water. If there has been previous heavy rain, the ground may be waterlogged – in this case the water would be unable to infiltrate and would travel overland to the river, reducing the lag time and increasing the peak discharge. The same would be true of a drainage basin in a very cold or a hot and dry climate as the ground would be hard and difficult for the water to permeate. A snowstorm will result in an increase in a river's discharge, but in this case there would likely be a long lag time due to the time taken for the snow to melt.

Vegetation

If the area surrounding the river has thick vegetation, it will have a long lag time and a low peak discharge. This is because much of the precipitation will be intercepted by the vegetation, which will then lose it through transpiration and evaporation.

Human activity

Urban areas are often covered in materials that water cannot infiltrate, such as tarmac. This will have the same effect as in a river surrounded by impermeable rock or soil; the water will travel overland, reducing the lag time and increasing the peak discharge. As water doesn't infiltrate easily in urban areas, humans often build storm drains that run directly into a river. This reduces the lag time and increases the river's peak discharge.

Figure 2.33: *Rivers in urban areas will have shorter lag times and higher peak discharges*

Usefulness of hydrographs

Analysis of hydrographs can help hydrologists predict the likelihood of flooding in a drainage basin as well as the possibility of a drought. It allows authorities to put in place flood defences in susceptible areas as well as reservoirs to store water.

Make the link

In the River basin management chapter you will see how hydrographs can be useful in determining how best to manage a river.

Summary

In this chapter you have learned:

- The processes involved in the formation of river features
- The formation of a V-shaped valley and a waterfall
- The formation of a meander and an oxbow lake
- The formation of erosional and depositional features in river landscapes
- The main components of the hydrological cycle
- The passage of water through the hydrological cycle
- The physical and human features affecting the hydrological cycle
- The definition of a river basin
- The main features of a drainage basin in relation to inputs, transfers and outputs
- How to describe a hydrograph
- The components of a hydrograph
- To analyse and interpret the data from a storm hydrograph
- How to explain the factors that affect a hydrograph
- How to compare hydrographs
- The purpose of hydrographs.

You should have developed skills and be able to:

- Evaluate the elements of the drainage basin
- Describe in detail the data from a hydrograph
- Analyse and evaluate data from a hydrograph
- Compare and evaluate data from different hydrographs.

End of chapter questions and activities

Quick questions

1. Explain the formation of a waterfall.

2. Explain the processes involved in the formation of a meander.

3. Explain why the transfer of water between land, air and sea can be described as a cycle.

4. Explain the distribution of water in the global hydrological cycle.

5. To what extent can human activity affect the global hydrological cycle?

6. Analyse the factors that affect the amount of water in a drainage basin.

7. With the aid of a diagram, explain the main features of a hydrograph. You should mention rising limb, receding limb, lag time and falling limb.

8. Look at Figure 2.34. Discuss how the features shown in the diagram affect a hydrograph.

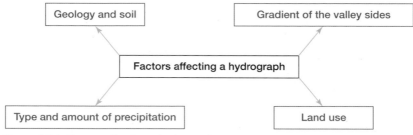

Figure 2.34: *Factors affecting a hydrograph*

9. Look at Figure 2.35. Compare these two hydrographs showing bare land and forested land. Which one has the largest lag time? Explain your answer.

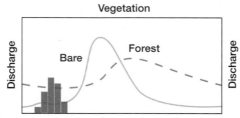

Figure 2.35: *Hydrograph showing bare land and forested land*

Exam-style questions

1. Study Figure 2.36

    a. Describe the changing river levels on the Rive Nene on 29 and 30 April.

    b. Explain the reasons for these changes.

    10 marks

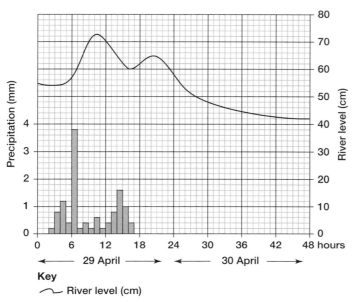

Figure 2.36: *Flood hydrograph for the River Nene in Northampton*

2. Study Figure 2.37.

    Explain how human activities, such as those shown in Figure 2.37, can impact on the hydrological cyle.

    8 marks

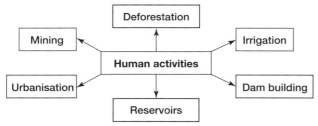

Figure 2.37: *Human activities affecting the hydrological cycle*

3. Look at Figure 2.38.

Meanders are found in the middle and lower course of rivers.

Explain the formation of a meander.

You may wish to use an annotated diagram or diagrams.   **8 marks**

Figure 2.38: *A meander*

Activity 1: Create a mind map

A drainage basin has inputs, outputs, storage and transfers.

In groups of four, complete a mind map for a river basin using the word bank below.

Each person should choose one component of the mind map. Expand the points.

Discuss with your group any other factors that affect a river basin, then add them to your mind map.

You have now created a good revision tool!

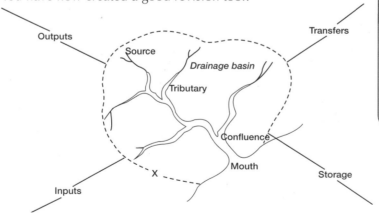

> 📖 **Word bank**
>
> Interception, Evaporation, Rain, Percolation, Groundwater, Vegetation, Lakes, Channel flow, Transpiration, Rivers, Snow, Reservoirs, Groundwater flow

Figure 2.39: *Mind map of a river basin*

Activity 2: Spider diagrams

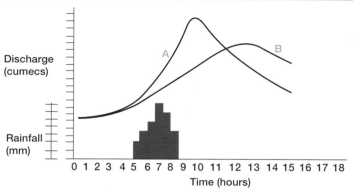

Figure 2.40: *Hydrographs A and B*

Hydrographs A and B are the result of the same storm but from different drainage basins.

a. With a partner, copy the two spider diagrams below. Complete the diagrams by matching the statements about the two drainage basins to either hydrograph A or hydrograph B.

Choose from:

- Urban area
- Steep slope
- Forestry plantation
- Sandy soil

- Rural area
- Gentle valley
- Impermeable rock
- Deforested area

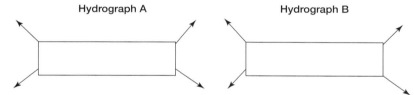

b. Using the information in the diagram and your completed spider diagram, explain the reasons for the differences in hydrographs A and B. Share your information with the rest of the class.

Activity 3: Weather conditions

The following weather conditions will affect a storm hydrograph by producing either a) a long lag time and gentle rising limb or b) a short lag time and steep rising limb. For each weather condition, select effect a) or b). Give reasons for your choice.

Weather conditions:

- A snowstorm in the Grampian mountains
- A thunderstorm in Glasgow
- A long period of drizzle during summer
- A rainstorm over the Queen Elizabeth Forest Park
- A rainstorm in an area of impermeable rock
- A rise in temperature after a heavy snowfall

Learning Checklist

You have now completed the Hydrosphere chapter. Complete a self-evaluation sheet to assess what you have understood. Use traffic lights to help you make up a revision plan to help you improve in the areas you have identified as amber or red.

- Explain the processes involved in the formation of river features.

- Explain the formation of erosional river features such as V-shaped valleys and waterfalls.

- Explain the formation of depositional river features such as meanders and oxbow lakes.

- Demonstrate an understanding of the circulation of water around the Earth.

- Explain the inputs, storage, processes and outputs of a drainage basin.

- Discuss the factors that affect a river basin.

- Discuss the main features of the hydrological cycle.

- Analyse the processes involved in the hydrological cycle.

- Explain the components of a hydrograph.

- Analyse a storm hydrograph.

- Compare two hydrographs.

- Discuss and evaluate the factors that affect a storm hydrograph.

- Suggest ways a hydrograph can be used.

Glossary

Base flow: the base flow of the river represents the normal day-to-day discharge of the river.

Channel: an area that contains flowing water confined by banks.

Condensation: the conversion of a vapour or gas to a liquid.

Confluence: where two rivers or streams meet.

Course: the route a river takes from source to mouth.

Current: the flow of the river.

Dam: a barrier built, usually across a watercourse, for holding back water or diverting the flow of water.

Deposition: the depositing of material carried from another part of the river.

Discharge: the amount of water carried by the river.

Drainage basin: the area drained by the river and its tributaries.

Erosion: the wearing away of river material.

Evaporation: the conversion of water from a liquid to a gas by solar energy.

Falling limb: downward trend of discharge line in a hydrograph.

Flood: when water breaks through the riverbanks and spreads over the surrounding land.

Freeze thaw: a crack in a rock fills with water, which then freezes when the temperature drops. Expansion and contraction over time weakens the rock, causing the crack to enlarge and the rock to break apart.

Hydrograph: graph that shows river discharge.

Infiltration: this is the process by which (water) precipitation soaks into the soil and moves through the cracks and pore spaces.

Irrigation: artificial watering of crops. In areas where there is not much rainfall, farmers irrigate the land by diverting water from rivers to their fields.

Lag time: the time difference between the peak precipitation and the peak discharge.

Lower course: the flatter area where a river normally enters the sea.

Middle course: the area between the upper course and the lower course where the river becomes wider and deeper and begins to meander.

Mouth: the point where the river enters the sea.

Peak discharge: the point on a flood hydrograph when river discharge is at its greatest.

Peak rainfall: the point on a flood hydrograph when rainfall is at its greatest.

Precipitation: a general term for all forms of water particles: rain, snow, sleet, dew, hail, etc.

Rising limb: the steep part of the discharge line in a hydrograph that has an upward trend.

Run-off: water that drains into a river.

Source: the starting point for a stream or river.

Storm hydrograph: a specific type of hydrograph that shows precipitation and discharge during and after a storm.

Sublimation: water loss from ice.

Through flow: the movement of water through the soil towards the river.

Transpiration: water loss from plants.

Transportation: the moving of river material from one place to another.

Tributary: a small stream or river that flows into a larger stream or river.

Upper course: the area where a river begins.

V-shaped valleys: steep-sided valleys usually found in the mountains and hills.

Water cycle: the natural cycle in which the Sun's energy evaporates water into the atmosphere, and the water vapour condenses, returning to the Earth as precipitation (rain, snow, sleet, etc.).

Watershed: an area of high land separating different river basins.

Water table: groundwater stored in permeable rocks.

3 Lithosphere

Within the context of the Lithosphere you should know and understand:

- Formation of erosion and depositional features in glaciated and coastal landscapes.

You also need to develop the following skills:

- Interpret maps and diagrams
- Use diagrams to explain the formation of landforms
- Evaluate photographs.

Scotland's landscape

The landscape of Scotland is changing all the time. The rocks that make up the landscape are, and have always been, attacked by the forces of nature. Today, these include the weather and rivers. In the past, it was glacial ice that shaped the Scottish landscape.

The effects of glaciation on the landscape

About two million years ago, our climate began to get colder. Rain turned to snow, which built up on hills and turned to ice. The **ice age** had begun. As more snow fell, the ice beneath was squeezed out and began to move slowly down the mountainside. This ice is called a **glacier**. Very soon, nearly all of Scotland was buried under hundreds of metres of ice. The extent of the last ice age is shown in Figure 3.1.

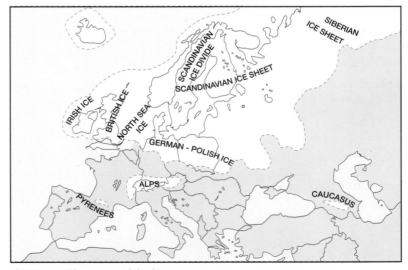

Figure 3.1: *The extent of the last ice age*

Glacial erosion

This ice carved and scraped away at the landscape using three key processes of erosion:

1. Freeze-thaw/frost-shattering

High in the hills and mountains, rain and snowfall are common. As the temperature falls overnight, water from rain and snow that has collected in cracks in the rock fast freezes and expands, causing the cracks to widen. When the temperature rises again, the water thaws and contracts. This process is repeated many times and pieces of rock are eventually broken off. These pieces of rock collect at the bottom of slopes as **scree**. This is shown in Figure 3.2.

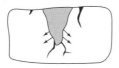

| Water fills a crack in a rock | The water freezes and the cracks widen | The rock breaks into several pieces |

Figure 3.2: *Freeze-thaw/frost-shattering*

2. Plucking

As a glacier is travelling slowly downhill over the landscape, rocks and stones from the floor and sides of the valley freeze to the glacier. The rocks are pulled from the ground as the glacier continues to move forward and these rocks and stones become incorporated into the base and sides of the glacier (see Figure 3.3).

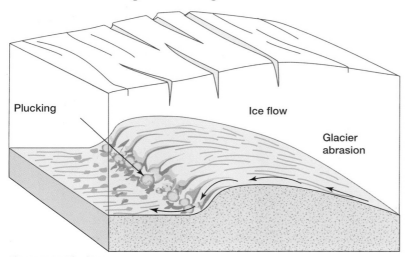

Figure 3.3: *Plucking*

3. Abrasion

This occurs when rocks and stones, picked up by the glacier through plucking, are rubbed against the bedrock at the bottom and side of the glacier (like sandpaper) as the glacier moves downhill. This causes wearing on the landscape. An indication of abrasion is the presence of **striations** cut into bare rock surfaces (see the photo in Figure 3.4). Striations are the result of fragments of rock in the glacier abrading the surface of rocks, leaving lines or gouges behind.

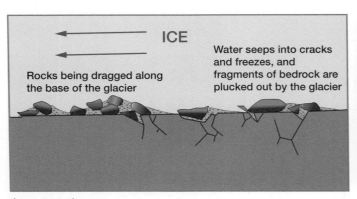
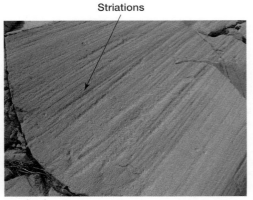

Figure 3.4: *Abrasion*

Formation of a corrie

Corries (from the Gaelic 'coire') are also known as cirques (French) and cwms (Welsh). A corrie is a steep-sided armchair-shaped hollow found high up in the hills or mountains.

Figure 3.5: *A corrie*

There are three stages in the formation of a corrie.

Stage 1: Before glaciation

1. Snow collects in sheltered hollows on the **north-facing** sides of mountains. The northerly aspect means greater protection from the Sun so that the snow never completely melts in the summer.

2. The snow fills up the hollow and compresses into ice. Over several years the snow builds up, squeezing out the air and creating granules of crystallised snow called **neve.**

3. The ice begins to move downhill, rotating slightly, due to gravity.

Stage 2: During glaciation

4. As the ice moves, it is plucking rocks and boulders from the **back wall** of the hollow, making it steeper and abrading the **sides and bottom** of the hollow, causing it to deepen.

5. Plucking and frost-shattering sharpen the back wall of the corrie and make it very steeply sloped.

Stage 3: After glaciation

6. Abrasion acts like sandpaper, making the hollow deeper.

7. When the ice melts, a small lake called a **lochan** or **tarn** is left.

8. The glacier deposits material it has gathered as a pile of moraine, forming a **lip** at the front edge and forming scree along the back wall.

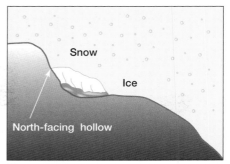

Figure 3.6: *Stage 1*

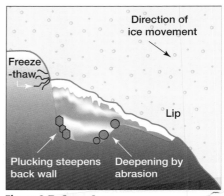

Figure 3.7: *Stage 2*

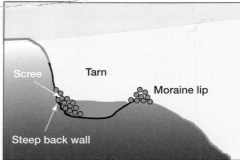

Figure 3.8: *Stage 3*

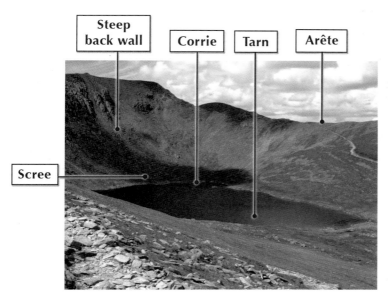

Figure 3.9: *Helvellyn*

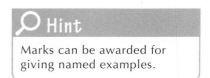

Hint

Marks can be awarded for giving named examples.

Corries are rarely found in isolation. Several usually form within the same mountainous area. When this happens, other glacial features called arêtes and pyramidal peaks form.

Formation of an arête

An arête is a thin, almost knife-like, ridge of rock high up in the mountains. Arêtes are formed in the following way:

1. When two corries formed back to back, narrow ridges called arêtes formed.
2. Frost-shattering further sharpened the ridge, leaving it rocky, jagged and often with a partial cover of scree.

Figure 3.10a: *Glaciated landscape*

Figure 3.10b: *An arête*

Hint

If asked for the formation of an arête or a pyramidal peak, first explain how the backwall of the corrie is formed and then discuss the formation of the arête/pyramidal peak.

Formation of a pyramidal peak

A pyramidal peak (or glacial horn) is a mountain top that has been sharpened by the action of ice during glaciation and frost-shattering. Pyramidal peaks are formed in the following way:

1. When three or more corries erode back to back around a mountain top, the arêtes between the corries rise to a central peak, called a pyramidal peak.
2. This is sharpened by frost-shattering.

Figure 3.11: *Pyramidal peak (left) and arête (right)*

Formation of a U-shaped valley

When glaciers are squeezed out of corries and move downhill under gravity, they tend to follow an existing river valley. Unlike a river, however, they fill the entire valley and their power to erode is much greater. This means that instead of having to wind around obstacles like a river, the glacier can follow a more direct route.

There are three stages in the formation of a U-shaped valley.

Stage 1: Before glaciation

1. The U-shaped valley begins as a **V-shaped** valley carved out by a river.

Stage 2: During glaciation

2. During the ice age, a glacier bulldozes its way down through this valley.

3. As the glacier flows through the valley, plucking removes rock from the valley sides and floor.

4. Rocks trapped in the ice act like sandpaper, wearing away the valley by abrasion.

5. The valley is greatly deepened, widened and straightened by plucking and abrasion.

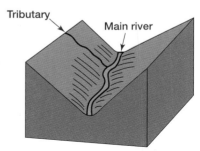

Figure 3.12: *Before glaciation*

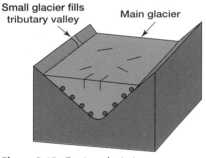

Figure 3.13: *During glaciation*

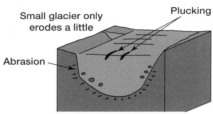

Figure 3.14: *Abrasion and plucking*

Stage 3: After glaciation

6. When the ice melts, the valley is **'U' shaped**.

7. It has very steep sides, a fairly flat floor and truncated spurs.

An example of a U-shaped valley is Glen Coe in Scotland. U-shaped valleys vary depending on rock hardness, glacier size and intensity of erosion but the features listed in Figure 3.15 are common.

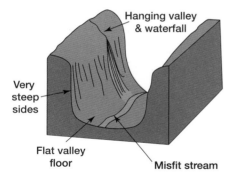

Figure 3.15: *After glaciation*

Figure 3.16: *U-shaped valley*

53

Hanging valley

Formation of a hanging valley:

1. A small glacier flows down from a corrie, eroding by **abrasion** to join the glacier in the main U-shaped valley.

2. This smaller glacier doesn't erode as deeply as the main valley glacier and after the ice melts a small **tributary valley** is left hanging above the main valley.

3. A small stream flows from the corrie and over the hanging valley, forming a **waterfall**.

4. At the bottom of the waterfall an **alluvial fan** is formed by deposition.

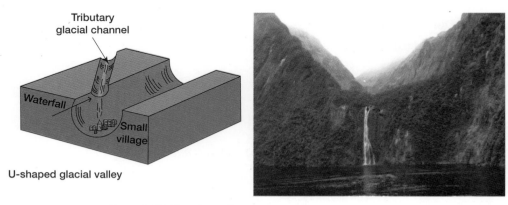

Figure 3.17: *Hanging valley*

Ribbon lakes and misfit streams

Formation of ribbon lakes and misfit streams:

1. Glaciers move down a former V-shaped valley and erode the valley by abrasion and plucking, turning the valley into a 'U' shape.

2. When temperatures rise and the glacier retreats, sometimes the meltwater forms a misfit stream.

3. This stream is too small to have eroded the U-shaped valley itself and so looks out of place on the valley floor; hence, it is called a misfit stream.

4. Some ribbon lakes form when the glacier meets softer rock.

5. The glacier erodes more deeply at these points.

6. After the glacier has melted, the deeper parts fill with meltwater to form ribbon lakes.

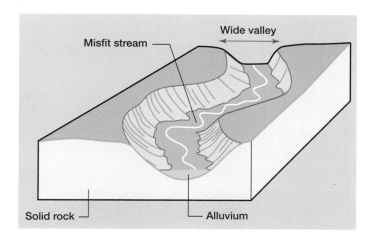

Side view of valley before glaciation   Side view of valley after glaciation

Figure 3.18: *Ribbon lakes and misfit streams*

Glacial deposition

When glacial ice melts, different types of rocks that have been carried by the glacier are deposited and piles of these deposits are called moraines. There are several different types of moraine:

1. **Medial moraines** are found in the middle, where two glaciers meet.
2. **Lateral moraines** are found deposited along the sides of the glacier.
3. **Terminal (end) moraines** are found at the terminus or the furthest (end) point reached by a glacier.

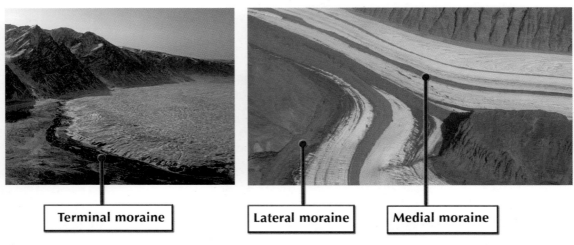

Terminal moraine   **Lateral moraine**   **Medial moraine**

Figure 3.19: *Moraines*

Formation of terminal moraine

Terminal moraine is a ridge of rocks, stones and debris (called boulder clay) found across the floor of a U-shaped valley.

Stage 1: During glaciation

1. The glacier bulldozed its way down through the valley, churning up rock and debris.
2. It pushed this forward in front of it at the glacier's **snout**.

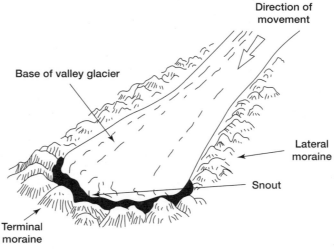

Figure 3.20a: *During glaciation*

Stage 2: After glaciation

3. When temperatures rose, the glacier began to melt and lose its power to transport material.

4. The unsorted material it had transported at its snout was then deposited as a ridge across the valley floor.

5. This ridge of terminal moraine marks the farthest point reached by the glacier (see Figure 3.20b).

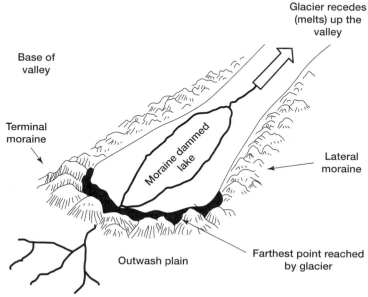

Figure 3.20b: *After glaciation*

Formation of drumlins

Drumlins are elongated hills of glacial deposit. They often occur in groups called 'a drumlin swarm' or 'a basket of eggs' such as in Vale of Eden in the Lake District. They measure up to 1 kilometre long and 500 metres wide.

Figure 3.21: *Drumlins*

Figure 3.22: *Side view of drumlin*

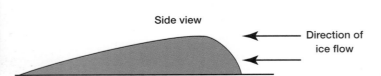

Side view

Direction of ice flow

The drumlin is made up of the debris that was carried along by a glacier and deposited when the glacier became overloaded with sediment. The long slope of the drumlin shows the direction in which the glacier was moving. The steep 'stoss' slope faces up the valley and the more gently sloping 'lee' slope faces down the valley. A small obstacle on the ground might act as a catalyst to allow till or boulder clay to build up around it.

Formation of eskers

Most eskers are believed to have formed in ice-walled tunnels by streams that flowed within and under glaciers. They are therefore made up of meltwater sands and gravels. These are sorted by size with large stones at the base because larger stones are dropped first by flowing water. The stones also tend to be more rounded than glacial deposits, because of the action of flowing water rounding the edges by erosion. As the glacier melts, sub-glacial streams flow through tunnels. After the retaining ice walls melted away, gravel and sand deposited by the streams remained as long winding ridges.

Figure 3.23: *Eskers*

Coasts

Coasts are a natural and constantly changing environment. Just like glaciated landscapes, they are shaped by processes of erosion, transportation and deposition.

Waves

Waves appear as a series of troughs and crests and vary considerably in height and length. The strength and height of a wave depends on the speed of the wind and the distance it has travelled. This distance is called the **fetch**. The largest fetch affecting the UK is caused by the south-westerly winds that bring waves from Brazil to the south coast of England.

Swash: the forward movement of a wave up a beach.

Backwash: the movement of water back out to sea when a wave has broken.

> ### ⚫ Make the link
>
> You may have learned about coastal erosion and deposition as part of the Coastal Landscapes topic in National 5 Geography.

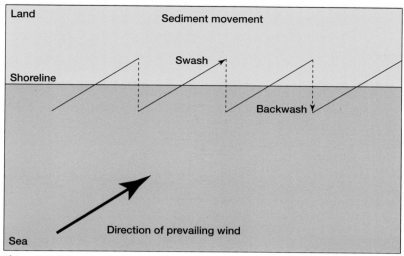

Figure 3.24: *Swash and backwash*

Constructive wave: a powerful wave with a strong swash that surges up a beach.

Destructive wave: a wave formed by a local storm that crashes down on a beach and has a powerful backwash.

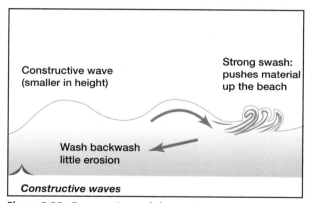
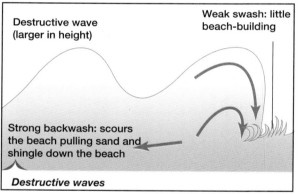

Figure 3.25: *Constructive and destructive waves*

Processes of coastal erosion

1. **Hydraulic action**

   Water and air is forced into cracks in the rocks of a coastline by waves. When the wave hits, pockets of air are compressed. When the wave retreats, the air expands quickly. This happens repeatedly and the cracks get larger until pieces of rock break off.

2. **Corrosion (also known as solution)**

   Soluble minerals from the rock are dissolved by the sea water. This type of erosion affects chalk cliffs in particular.

3. **Corrasion (also called abrasion)**

   This is the erosion of the base of a cliff that happens as sand and rocks are thrown at the cliff face by waves. These scrape away the rock over time, undercutting the base of the cliff.

4. Attrition

Rocks that have been broken away from the coast by one of the processes above collide with one another and the rock face in the waves, becoming smaller and rounder.

Formation of cliffs and wave-cut platforms

Sea cliffs are the most common landform of coastal erosion. Cliffs begin to form when destructive waves attack the base of the rock.

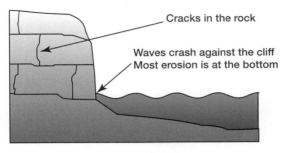
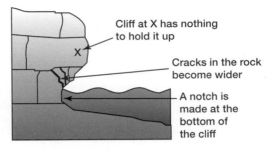

Figure 3.26: *Formation of cliffs and wave-cut platforms*

Processes such as hydraulic action and corrasion undercut the base of a cliff to form a wave-cut notch. The rock above overhangs over the notch and, as the waves continue to attack, the notch widens until the overhang is no longer supported and collapses due to the weight.

The waves gradually remove the fallen rock and then begin to attack the new cliff face again. Over time the **cliff retreats inland** leaving a **wave-cut platform** (flat land at the foot of the cliff).

Figure 3.27: *Wave-cut notch*

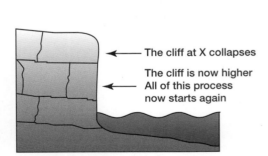

Figure 3.28: *Wave-cut platform*

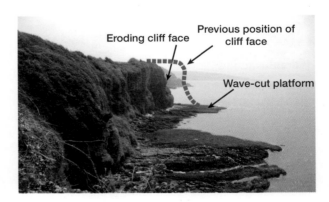

Formation of headlands and bays

A headland is an area of hard rock that juts out into the sea. A bay is a sheltered area between headlands.

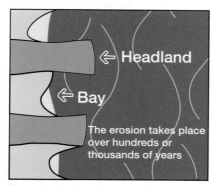

Figure 3.29: *Headlands and bays*

Figure 3.30: *Swanage Bay*

Headlands form along coastlines in which bands of **soft and hard rock** outcrop at **right angles** to the coastline. **Differential erosion** occurs where the soft, less resistant rock (e.g. shale) erodes more quickly than the hard, resistant rock (e.g. chalk).

🔍 Hint

In your exam, it is a good idea to include named examples when explaining the formation of a feature, such as Swanage Bay. These are easy to remember and including them demonstrates good knowledge.

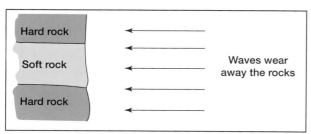

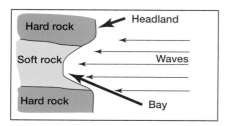

Figure 3.31: *Formation of headlands and bays*

The soft rock is eroded more rapidly (by corrasion, corrosion and attrition) than the hard rock, forming bays. The more resistant rock is eroded at a slower rate by hydraulic action, leaving the hard rock sticking out into the sea as a headland. The exposed headland now becomes vulnerable to the force of destructive waves but shelters the adjacent bays from further erosion.

Formation of caves, arches, stacks and stumps

Waves widen any weakness or crack in a rock by the process of **hydraulic action**. This can result in the formation of caves on a headland.

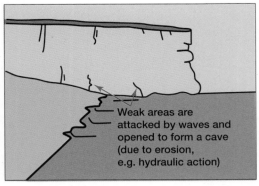

Weak areas are attacked by waves and opened to form a cave (due to erosion, e.g. hydraulic action)

Figure 3.32: *Formation of caves; Tilly Whim Caves (Dorset)*

Over time, vertical erosion may cause a blowhole to form in the roof of the cave. Waves can continue to erode inside the cave through corrasion until eventually they break through the 'back wall' to form an **arch**. Corrasion and corrosion continue, attacking the base of the arch and putting pressure on its roof.

Cave widened and deepened by erosion to form an arch

Figure 3.33: *Formation of arches; Durdle Door (Dorset)*

Eventually, over time the roof will collapse due to continued erosion and chemical weathering, leaving a tall isolated **stack**. The force of the waves attacks the stack at the base (undercutting) and eventually it will collapse to form a **stump**.

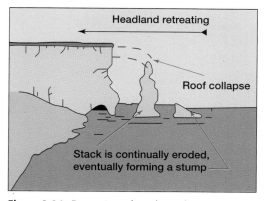 

Figure 3.34: *Formation of stacks and stumps; Old Harry (Dorset)*

Figure 3.35: *Old Harry and his wife*

? **Did you know?**

At one time, Old Harry had a wife – another stack that was eroded and eventually collapsed into the sea about 50 years ago! You can see her remains when the tide is low.

Coastal deposition

Formation of a beach

The transport of sand and pebbles along the coast is called longshore drift. The swash, carrying sand and pebbles, approaches the coast at an angle according to the direction of the prevailing wind. The backwash carries the material back down the beach at a right angle to the coastline under gravity, meaning sand and pebbles in the water are carried along the coastline in a zig-zag pattern. In the diagram below, the prevailing wind is approaching from the south-west and therefore longshore drift is moving material from the west to the east. This material will be deposited on the beach when the waves lose energy.

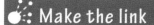 **Make the link**

You may have learned more about the effects of wind when learning about weather for National 5 Geography.

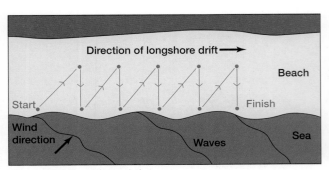

Figure 3.36: *Longshore drift*

Formation of a sand spit

Longshore drift moves material such as sand and shingle along the coast. Material is moved by the action of the waves (swash) in the direction of the prevailing wind. It is then carried back down the beach at right angles by the backwash. Where the **coastline changes direction**, material continues to be deposited along the original direction, even though there is not a coastline to follow. Deposited material builds up and eventually breaks through the surface. Sometimes a **hook** develops if the wind changes direction. Waves cannot get beyond the spit, creating a sheltered area where mud flats or **salt marshes** form.

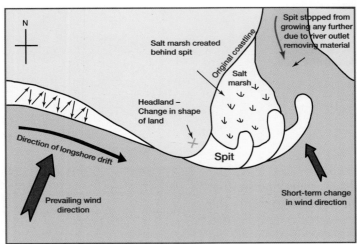

> ### 🔍 Hint
>
> Practise drawing the above diagram – if you are asked about longshore drift in the exam, a well-labelled sketch will help you explain clearly.

Figure 3.37: *Formation of a spit; Spurn Head (Yorkshire)*

Formation of a sand bar

Where a spit joins one headland to another it is called a **sand bar**. It is a barrier of sand stretching right across a sheltered bay. It can cut off a marsh or **lagoon** to the landward side, e.g. Slapton Sands in Devon.

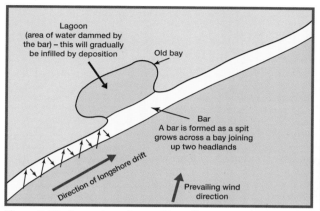 

Figure 3.38: *Formation of a sand bar; Slapton Sands (Devon)*

Formation of a tombolo

Where a spit joins a headland to an island it is called a **tombolo**. It is a barrier of sand stretching out to sea where it meets land. The longest and best-known tombolo in Britain is **Chesil Beach** (Dorset).

> 🔍 **Hint**
>
> If asked for the formation of a sand spit, bar or tombolo, first explain in detail the process of longshore drift.

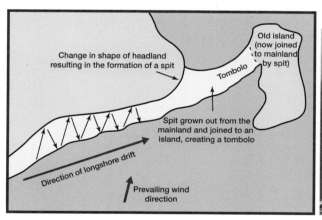 

Figure 3.39: *Formation of a tombolo; Chesil Beach (Dorset)*

Summary

In this chapter you have learned:

Glaciation

- The processes of glacial erosion
- The formation of features of glacial erosion including corries, arêtes, pyramidal peaks, U-shaped valleys, hanging valleys, ribbon lakes and misfit streams
- The formation of features of glacial deposition including terminal moraine, drumlins and eskers.

Coasts

- The processes involved in coastal erosion
- The formation of features of coastal erosion including wave-cut platforms, headlands and bays, caves, arches, stacks, stumps
- The formation of features of coastal deposition including sand spits, sand bars and tombolos.

You should have developed skills and be able to:

- Draw and annotate diagrams showing formation of glacial and coastal features
- Identify glacial and coastal features from diagrams.

End of chapter questions and activities

Quick questions

Glaciation

1. Explain the processes of freeze-thaw, plucking and abrasion.

2. Identify the glacial features shown on Figure 3.40.

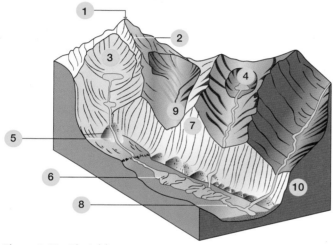

Figure 3.40: *Glacial features*

3. Identify and describe the features that show that the area in Figure 3.41 has been glaciated.

Figure 3.41: *Glaciated landscape*

4. Explain, with the aid of diagrams, the formation of one feature of glacial erosion and one feature of glacial deposition.

Coasts

5. Describe and explain the movement of waves.

6. Explain the four processes of coastal erosion.

7. Identify the coastal features shown on Figure 3.42.

Figure 3.42: *Coastal landscape*

8. Explain with the aid of a diagram(s) the formation of headlands and bays.

9. Explain longshore drift.

Exam-style questions

1. Study Figure 3.43.

   Choose either an erosional feature or a depositional feature shown on Figure 3.43. With the aid of annotated diagrams, **explain** the formation of your chosen feature.

   8 marks

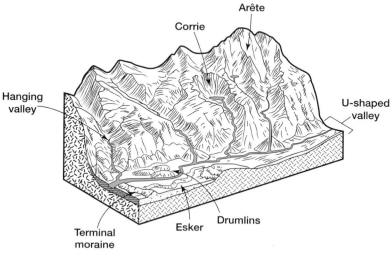

Figure 3.43: *Glaciated landscape*

2. Look at Figure 3.44.

   Explain the formation of a sand spit.

   You may wish to use an annotated diagram or diagrams in your answer.

   8 marks

Figure 3.44: *Sand spit*

Activity 1: Stop-motion animation

Begin by creating a storyboard illustration of how one feature of coastal erosion or deposition is formed. Once you have completed this, use a digital camera and tripod, some Play-Doh and a movie-making software app to produce a stop-motion animation of your storyboard.

First, put the camera on the tripod and place the Play-Doh on your table in the first position on your storyboard. Take a picture, then move the Play-Doh slightly into the second position and take another photo. Repeat until you have plenty of frames. Ten frames will only last around 3 seconds so it will take time and patience to build up your animation. Upload the images onto your computer and use a movie-making software app to string them together into one clip. Share your videos with your class!

Activity 2: What am I?

Each person in your class will be asked to write down a description of a feature or process of glacial erosion or deposition. Each class member will then take it in turns to read their description to the class. The class will have 30 seconds to write down what they think is being described. Have a partner peer assess your answers once all descriptions have been read out.

Learning Checklist

You have now completed the Lithosphere chapter. Complete a self-evaluation sheet to assess what you have understood. Use traffic lights to help you make up a revision plan to help you improve in the areas you have identified as amber or red.

Glaciation

- Explain the formation of a glacier.

- Identify on a map the extent of the ice age in Europe.

- Identify glacial features from a diagram.

- Explain the processes of glacial erosion:

  ➢ plucking

  ➢ abrasion

  ➢ freeze-thaw.

- Explain with the aid of a diagram the formation of erosional features such as:

  ➢ corries

  ➢ arêtes

  ➢ pyramidal peaks

  ➢ U-shaped valleys

  ➢ hanging valleys.

- Explain the formation of depositional features such as:

  ➢ terminal moraine

  ➢ eskers

  ➢ drumlins.

(continued)

Coasts

- Identify coastal features from a diagram.

- Explain the processes of coastal erosion including:

    ➢ hydraulic action

    ➢ corrosion

    ➢ corrasion

    ➢ attrition.

- Explain with the aid of a diagram the formation of coastal erosional features such as:

    ➢ cliffs and wave-cut platforms

    ➢ headlands and bays

    ➢ caves, arches and stacks.

- Explain the formation of depositional features such as:

    ➢ sand spits

    ➢ sand bars

    ➢ tombolos.

Glossary

Glaciation

Ablation: the loss of ice and snow from a glacier.

Abrasion: erosion caused by rocks and boulders in the base of the glacier acting like sandpaper, scratching and scraping the rocks below.

Accumulation: the addition of snow or ice to a glacier.

Advance: an increase in the length of a glacier compared to previous times.

Alluvial fan: a fan- or cone-shaped area of sediment ('alluvium') that is deposited when a stream or river reaches a level surface at the bottom of a slope.

Arête: sharp, knife-like ridge formed between two corries cutting back.

Aspect: the direction a corrie faces.

Corrie: a bowl-shaped hollow area formed by glaciation. Corries sometimes have lakes in them and they are also known as cirques or, when found in Wales, cwms.

Crevasse: a deep crack on the surface of an ice sheet or valley glacier.

Drumlin: an oval-shaped hill formed from deposits within a glacier.

Erosion: the wearing away of the land by rivers, ice sheets, waves and wind.

Eskers: long ridges of sand and gravel deposited by rivers that flowed under an ice sheet.

Freeze-thaw or frost-shattering: rock fragments break off due to water entering cracks in the rock expanding and contracting due to temperature change.

Glaciation: the build-up of ice on the land during colder periods in time.

Glacier: a slow-moving ice mass, formed over a long period from compacted snow.

Ground moraine: deposits at the base of the glacier, a result of abrasion and plucking of the valley floor.

Hanging valley: a tributary valley affected less by glaciation and left hanging above the main valley.

Ice age: a period of colder climate when ice sheets form on the land, causing a lowering of sea level.

Lateral moraine: rock debris that runs along the sides of a glacier resulting from ice erosion of the valley sides and freeze-thaw weathering on the bare rock above.

Medial moraine: rock debris that runs down the centre of the glacier. It forms from the merging of the lateral moraines of two glaciers.

Moraines: frost-shattered rock debris and material eroded from the valley floor and sides, transported and deposited by glaciers.

Misfit stream: a stream left after the ice has melted that is too small for the valley it occupies.

Plucking: as the ice moves forward it plucks or pulls out large pieces of rock.

Pyramidal peak: a jagged mountain summit where three or more corries form back to back.

Ribbon lakes: long, narrow lakes found in glaciated valleys formed in locations where the glacier had more erosive power.

Rotational movement: avalanches of snow collecting at the back wall of a corrie apply great pressure, forcing the ice out of the front of the hollow in a rotational movement.

Scree: a slope of loose, large, angular rocks broken away from the mountainside by freeze-thaw weathering.

Snout: the end of the glacier where melting occurs.

Snowline: the height at which permanent snow begins in mountainous regions.

Terminal moraine: a ridge of debris dumped at the end of a glacier and formed of unsorted boulders, sand, gravel and clay.

U-shaped valley: a river valley widened and deepened by the action of glaciers.

V-shaped valley: steep-sided valley usually found in the mountains and hills.

Coasts

Abrasion: rocks carried by the waves erode the coastline when they are thrown against headlands by the force of the waves.

Arch: opening through a small headland caused by two caves eroding back towards each other until the back walls disappear.

Attrition: the wearing down of the load as the rocks and pebbles hit the river bed and each other, breaking into smaller and more rounded pieces.

Backwash: the return of water to the sea after waves break on a beach.

Bay: a sheltered area between headlands.

Beach: deposition of sand and shingle along the coastline.

Blowhole: feature caused by the roof of a cave collapsing.

Cliff: a vertical rock face.

Coastline: the area where the land meets the sea.

Constructive wave: a powerful wave with a strong swash that surges up a beach.

Corrasion: when rocks carried by water wear away the landscape.

Corrosion: when chemicals in the water dissolve minerals in the rocks, causing them to break up.

Depositional feature: a landform that has been formed through the dumping of beach material.

Destructive wave: a wave formed by a local storm that crashes down on a beach and has a powerful backwash.

Differential erosion: where soft, less resistant rock (e.g. shale) erodes quicker than hard, resistant rock (e.g. chalk).

Erosion: the wearing away of the landscape.

Fetch: the distance a wave travels.

Headland: an area of hard rock that juts out into the sea.

Hydraulic action: erosion caused by waves hitting the cracks on a cliff face.

Lagoon: a shallow area of water separated from the sea by a sand bar.

Line of weakness: a part of a rock that is more susceptible to erosion.

Longshore drift: the movement of material such as sand and shingle along the coast.

Pebble: small, round rock.

Resistant rock: hard rock, more resistant to erosion.

Sand bar: where a spit joins one headland to another.

Spit: a stretch of beach at one end of a coastline caused by waves depositing material.

Stack: a tall piece of rock separated from the headland.

Stump: a small piece of rock left after erosion of a stack.

Swash: the forward movement of a wave up a beach.

Tombolo: a feature formed where a spit joins a headland to an island.

Undercutting: the wearing away of the base of a cliff.

Wave-cut platform: flat land at the foot of the cliff.

4 Biosphere

Within the context of the Biosphere you should know and understand:

• Properties and formation processes of podzol, brown earth and gley soil.

You also need to develop the following skills:

• Interpret a wide range of graphical information
• Analyse and synthesise information from soil profiles.

'Essentially, all life depends upon the soil [...] There can be no life without soil and no soil without life; they have evolved together.'

Charles E. Kellogg, USDA Yearbook of Agriculture, 1938

Our planet is the only one in our solar system to have life! All forms of life are dependent to some degree on soil and without it humans would not be able to grow crops or raise animals. It is our primary resource. The study of soil is called pedology, and there are over 50 different types of soil, depending on the conditions in an area.

How is soil formed?

The process of soil formation is slow. It can take up to 1000 years to form a soil one centimetre in depth. Every soil develops from parent material found on the Earth's surface. This bedrock can be weathered in place or it can be moved by glaciers, rivers or winds. Over time, sun, water, wind, ice and living creatures change the parent material into soil.

Soils differ from one part of the world to another and even from one garden to another. They differ because of where and how they formed.

Studies of soils show that the development of a soil is controlled by six major factors:

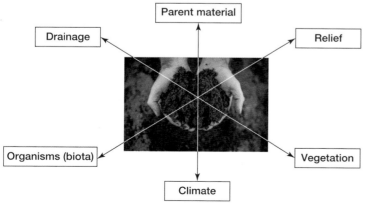

Figure 4.1: *Development of a soil*

🔍 Hint

A good way to remember these is to use a mnemonic, for example:

Vain Octopuses Really Can Polka Dance
V = Vegetation
O = Organisms (biota)
R = Relief
C = Climate
P = Parent material
D = Drainage

Figure 4.2: *Factors affecting the formation of soil*

1. Parent material

The parent material is very important to the type of soil formed. Limestone and chalk form alkaline soils, whereas materials like sandstone and clay will form acidic soils. Hard rock will take longer to break down into soil and will also tend to form coarser, grittier soils than softer rocks.

2. Climate

Temperature and precipitation play a big role in soil formation. In cold climates, the temperature and ice can aid the break-up of parent material (the ice may pluck or abrade the rock). However, heat and humidity can also speed up the process. The temperature will also affect the organisms and vegetation. In particular, if it is very cold then there will be fewer organisms working within a soil. In areas where there is high rainfall, minerals in the soil can be washed or leached out of the soil (see Figure 4.3).

Figure 4.3: *A rainy climate affects the quality of soil*

3. Relief and drainage

Altitude has an effect on soil formation: higher up there is more precipitation and it is colder. Conversely, it tends to be drier and warmer closer to sea level. If the formation is occurring on a slope, the direction of the slope will also affect how cold and wet it is; for example, west- or south-facing hills will usually be warmer and wetter, leading to the formation of richer soils. On a slope, the faster run-off of water will lead to greater erosion and the formation of thinner, less fertile soils. This soil may wash down and accumulate on flat land at the bottom of the slope, where there will likely be more leaching and waterlogging.

Figure 4.4: *Relief affects soil formation*

4. Organisms

Organisms such as worms, wood lice, moles, spiders, fungi and bacteria all help to make soil by breaking down organic and inorganic matter. The number of organisms is affected by the other factors in this list (as you can see, all the factors are inter-connected). They feed on both inorganic and organic matter (dead and alive) and also contribute nutrients themselves in the form of waste or dead bodies. Small animals and insects also mix the soil. In areas with many organisms of different types, healthy soils develop relatively quickly. Humans can also affect the formation of soil, not always for its improvement.

Make the link

In the River basin management chapter you will learn more about managing rivers and the effects this can have on the local environment.

Figure 4.5: *Worms and other organisms play an important role in soil formation*

5. Vegetation

Soil and the vegetation that grows on it are very closely linked; the vegetation supplies the soil with organic matter and the type of vegetation will have an effect on the type of soil; for example, in areas of coniferous forest the pine needles create an acid soil. In addition to this, the roots of some plants help to break down the parent material, affecting the soil formation. As the vegetation dies, the soil converts its remains into nutrients, which in turn helps vegetation to develop and thrive in the future.

Figure 4.6: *In a coniferous forest, the pine needles create an acid soil*

Podzol, brown earth and gley soils

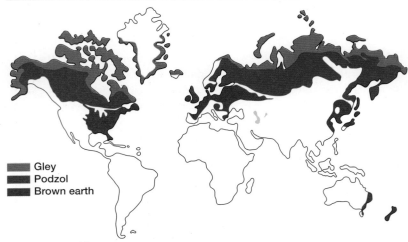

Gley
Podzol
Brown earth

Figure 4.7: *Soil locations*

We are going to look at three different soils in this chapter – podzol, brown earth and gley. Figure 4.7 is a map showing you where in the world each of these three soils is found.

Factors affecting the formation of podzol, brown earth and gley soils

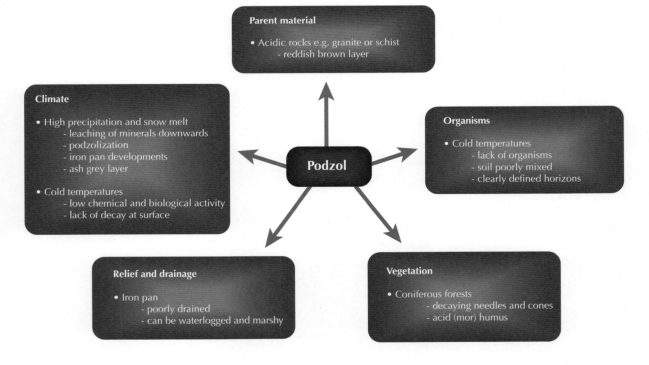

Parent material

- Acidic rocks e.g. granite or schist
  - reddish brown layer

Climate

- High precipitation and snow melt
  - leaching of minerals downwards
  - podzolization
  - iron pan developments
  - ash grey layer

- Cold temperatures
  - low chemical and biological activity
  - lack of decay at surface

Podzol

Organisms

- Cold temperatures
  - lack of organisms
  - soil poorly mixed
  - clearly defined horizons

Relief and drainage

- Iron pan
  - poorly drained
  - can be waterlogged and marshy

Vegetation

- Coniferous forests
  - decaying needles and cones
  - acid (mor) humus

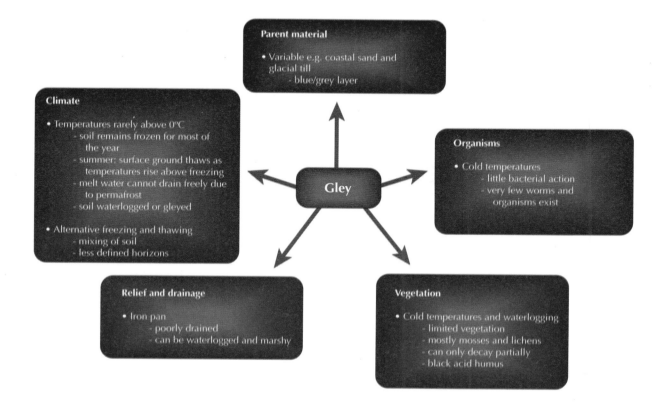

Brown earth

Parent material
- Variable: acidic/alkaline e.g. sandstone, limestone and bolder clay

Climate
- Precipitation lower than podzol
  - less leaching
  - less chance of iron pan developing
  - stronger colour
- Temperatures warmer than podzol
  - more chemical and biological activity
  - more decay at surface

Organisms
- Warmer temperatures
  - many worms and other organisms
  - mixed soil
  - less defined horizon

Relief and drainage
- Better drainage than podzol
  - less waterlogging and marsh
  - more fertile than podzol

Vegetation
- Deciduous forests
  - diverse selection of plants and trees e.g. oak, beech, fir and chestnut
  - mild acid (mull) humus

Gley

Parent material
- Variable e.g. coastal sand and glacial till
  - blue/grey layer

Climate
- Temperatures rarely above 0°C
  - soil remains frozen for most of the year
  - summer: surface ground thaws as temperatures rise above freezing
  - melt water cannot drain freely due to permafrost
  - soil waterlogged or gleyed
- Alternative freezing and thawing
  - mixing of soil
  - less defined horizons

Organisms
- Cold temperatures
  - little bacterial action
  - very few worms and organisms exist

Relief and drainage
- Iron pan
  - poorly drained
  - can be waterlogged and marshy

Vegetation
- Cold temperatures and waterlogging
  - limited vegetation
  - mostly mosses and lichens
  - can only decay partially
  - black acid humus

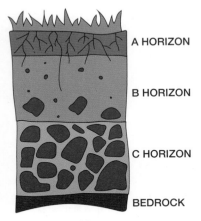

Figure 4.8: *Soil horizons*

Figure 4.9: *A podzol soil*

Soil profiles

There is a thin layer of soil over most of the Earth's land. A soil profile is a vertical cross-section that shows the different layers, which are called 'horizons' (see Figure 4.8). Underneath the surface layer (which is usually partly decayed vegetation and leaves) there are usually three main horizons – the A, B and C horizons.

Podzol soil

These soils are found in the higher, wetter areas of northern and western Britain. They tend to be found on the upper slopes of upland areas where precipitation is heavy and are associated with coniferous forest. The pine needles from the coniferous trees form a thin, acid layer and due to lower temperatures take a long time to decay. The slow rate of weathering of the parent rock results in a shallow soil that is usually infertile due to its acidity and lack of humus. There is little evapotranspiration due to the low temperatures, and this coupled with the rainfall results in iron and other minerals being leached downwards to leave the A horizon an ash-grey colour. Where the iron is deposited, it forms a rust-coloured hard pan, which limits drainage. Podzols are easily recognisable by their distinct layers or horizons, which mean that nutrients are recycled more slowly; this is due in part to the fact that it is cold for organisms, and so the soil is relatively unmixed.

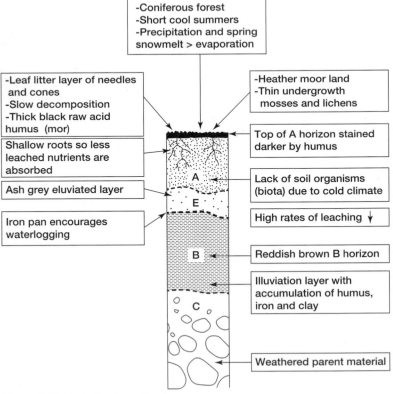

Figure 4.10: *Podzol soil profile*

Brown earth

Brown earth develops on lowland areas in southern Britain. Chemical and biological weathering break down the parent material quickly, resulting in a relatively deep and fertile soil. It is especially prevalent in areas of deciduous woodland and thick undergrowth; here, lots of leaves fall and decay rapidly in the autumn, producing a thick litter layer that is quickly broken down by worms and other organisms. This, alongside the quick recycling of nutrients, results in a dark brown upper horizon. Water moves down through the horizons slowly due to the fact that there is more precipitation than evapotranspiration causing some leaching and this, alongside the mixing action of the organisms, means that there is no distinct boundary between the A and B horizons. Iron and other minerals are leached downwards, giving the lower layer a reddish-brown tint.

Hint

A fully annotated diagram can gain full marks if detailed enough.

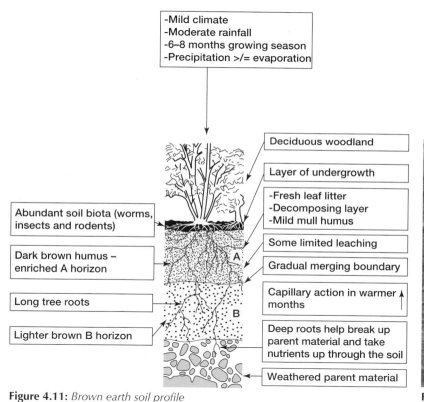

Figure 4.11: *Brown earth soil profile*

Figure 4.12: *A brown earth soil*

Gley

Gley soil is the most widespread type of soil in Britain and is typically found at the foot of slopes and in floodplains. Gley soils develop in sites that are waterlogged (either permanently or temporarily), which slows down the decay of bacteria. This causes the A horizon to be dark as there is a large quantity of organic matter that is not being broken into humus by bacterial activity. The B horizon is mainly blue-grey due to the waterlogging, which chemically changes minerals in the soil such as iron from a (normally) red-brown to a blue-grey colour (when the soil dries out, this process is reversed). The C horizon is derived from an impermeable clay layer.

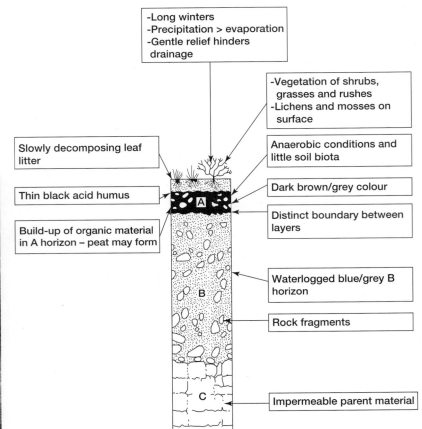

Figure 4.13: *A gley soil*   **Figure 4.14:** *Gley soil profile*

Diagram labels:
- Long winters
- Precipitation > evaporation
- Gentle relief hinders drainage
- Vegetation of shrubs, grasses and rushes
- Lichens and mosses on surface
- Slowly decomposing leaf litter
- Anaerobic conditions and little soil biota
- Thin black acid humus
- Dark brown/grey colour
- Distinct boundary between layers
- Build-up of organic material in A horizon – peat may form
- Waterlogged blue/grey B horizon
- Rock fragments
- Impermeable parent material

Summary

In this chapter you have learned:

- What soil is
- Why different soils form in different areas of the world
- The factors affecting soil formation
- The factors affecting the formation of a podzol
- The factors affecting the formation of a brown earth soil
- The factors affecting the formation of a gley soil
- To identify and explain the soil profile of a podzol
- To identify and explain the soil profile of a brown earth
- To identify and explain the soil profile of a gley soil.

You should have developed skills and be able to:

- Draw and annotate a soil profile
- Interpret a soil profile
- Identify the location of podzols, brown earths and gleys from a map.

End of chapter questions and activities

Quick questions

1. Explain the main factors that affect soil formation.

2. Using Figure 4.15, describe the locations of podzols, brown earth and gley soils.

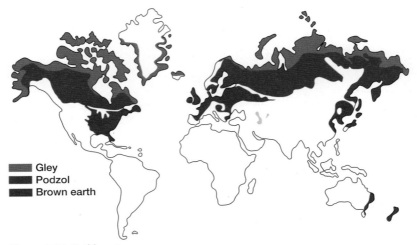

■ Gley
■ Podzol
■ Brown earth

Figure 4.15: *Soil locations*

3. Explain the reasons why brown earths and gley soils have developed in these areas.

4. For a podzol soil, explain the main conditions and soil-forming processes that have led to its formation.

5. Look at Figure 4.16. Explain how the relief in this photograph can affect soil formation.

Figure 4.16: *Relief*

6. Look at Figure 4.17. What kind of soil is likely to form in this environment? Give reasons to support your answer.

Figure 4.17: *Forested area*

Exam-style questions

1. Study Figure 4.18.

   Explain how factors such as natural vegetation, soil organisms, climate, relief and drainage have contributed to the formation and characteristics of either a podzol or gley soil.

   8 marks

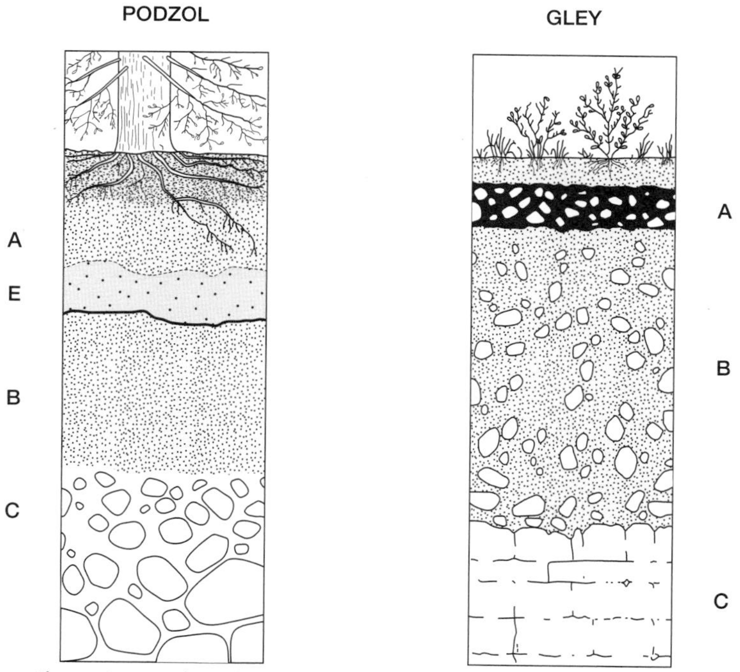

PODZOL

GLEY

A
E
B
C

A
B
C

Figure 4.18: *Selected soil profiles*

2. **Draw and fully annotate** a soil profile of a brown earth soil to show its main characteristics (including horizons, colour, texture, soil biota and drainage) and associated vegetation.

   10 marks

Activity 1: Peer teaching – teach the class about soil

Divide into groups. Select a soil type.

Create a poster, PowerPoint or podcast showing the main facts about your soil. Your poster/PowerPoint/podcast should include a location map, general characteristics about your soil, soil forming factors, soil profile and interesting facts.

Use your poster/PowerPoint/podcast to teach the rest of the class about your soil. Each group member should present part of the group findings to the class – plan your presentation! Your poster should be eye-catching and your presentation interesting. Be prepared for questions at the end!

Activity 2: Revision cards

In pairs, collect small pieces of card and make up revision cards. Use the glossary for this chapter and write each word on the front of the card and its definition on the back. Test each other to see if you can explain each word.

Learning Checklist

You have now completed the Biosphere chapter. Complete a self-evaluation sheet to assess what you have understood. Use traffic lights to help you make up a revision plan to help you improve in the areas you have identified as amber or red.

- Demonstrate an understanding of soil formation.

- Explain why different soils form in different areas.

- Locate and identify on a map the areas of the world where podzols, brown earths and gley soils are found.

- Discuss the main factors affecting the formation of a podzol.

- Discuss the main factors affecting the formation of a brown earth soil.

- Discuss the main factors affecting the formation of a gley soil.

- Draw and annotate the soil profile for a podzol, brown earth and gley soil.

- Be able to identify the three soil profiles.

Glossary

Bacteria: microscopic organisms that help break down a soil.

Bedrock: the solid unweathered rock that lies beneath all soil or other loose material.

Capillary action: the upward movement of water through a soil.

Climate: average weather conditions over at least 35 years.

Eluviation: downward movement of material where rainfall exceeds evaporation.

Hard pan: a hardened layer of soil caused by leaching of soils.

Horizon: a distinct layer of soil found in vertical section.

Humus: decomposed dead organic material.

Illuviation: accumulation of dissolved or suspended soil materials in one area or layer.

Leaching: the process whereby soil nutrients are removed by the action of water.

Parent material: weathered bedrock forming the basis of a soil.

Percolation: the downward movement of water through the soil.

Soil biota: creatures that live in the soil.

Soil profile: a vertical section through a soil.

Topsoil: the upper part of the soil, which is the most favourable material for plant growth.

Human Environments

5 Population

Census

The term **population** refers to the total number of people inhabiting a specified area. Nearly all countries count their population in some way or other. The aim of this is to gather information that can then be used for planning resources, education, employment, housing and other government services. Examples of how this information could be used include in planning the location of schools or care homes for the elderly.

Figure 5.1: *Crowds of people in Edinburgh*

One method of gathering this information is called a **census**. This is a type of questionnaire that every household is required by law to complete. In Scotland, a census is carried out every 10 years – the last one took place in 2011 and the next will take place in 2021.

Every population's characteristics are always changing. Census statistics help paint a picture of this and how we live. They provide a detailed 'snapshot' of the population and help government bodies decide how best to spend the country's money.

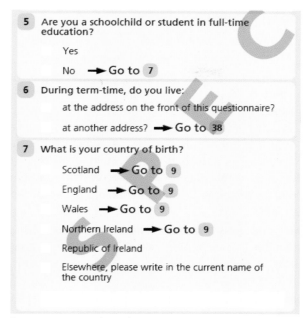

Figure 5.2: *Extract from the census*

Apart from the census, in many countries people are also required to register important events in their lives – these are called **vital registrations**. Vital registration is the method by which a government records the vital events of its population. Vital registrations in the UK include a record of births, deaths, marriages, divorces and adoptions. This registration of important life events, such as births, deaths and marriages results in the production of legal documents. For example, when a baby is born it must be registered within 42 days in England, Wales and Northern Ireland and within 21 days in Scotland. The baby is then issued with a birth certificate.

Births, deaths and marriages are recorded officially so that the government has a continually updated picture of the changing make-up of the population. This allows the government to change allocated funds to different areas.

Household surveys are carried out at various times. These are random and address-based and residents are asked to supply information on age, education, housing and income.

Border agencies also gather population data by requesting information from people entering the country.

Figure 5.3: *A couple with their marriage certificate*

Hint

Make sure you can discuss the problems of census taking in both developed and developing countries.

? Did you know?

It is estimated that approximately 800 languages are used in India

Make the link

If you take Modern Studies, you may learn more about social inequalities that contribute to these difficulties in both the UK and other countries around the world.

Figure 5.4: *There were protests in Burma in 2014 when the government refused to let people from the Rohingya Muslim minority identify themselves as such in the census*

Problems of taking a census

Conducting a census in both developed and developing countries can be difficult. Some of these difficulties are shown below.

1. They are very **expensive and time consuming** (forms have to be **produced**, **distributed,** **collected** and **collated**) and this is a significant problem, especially for a poorer developing country.

2. **Lower literacy rates in developing countries**, especially amongst women, mean that it is impossible for some people to fill in the census form as they cannot read or write, making the census inaccurate. Extra enumerators may help to overcome this problem but this will add to the expense. Ethiopia, for example, has literacy rates of less than 50%.

3. **Difficult terrain and poor communications** can also be a hindrance, especially in developing countries. If remote areas (mountain villages in Nepal for example) are not included in a count, it will be inaccurate.

4. In some areas **many different languages** are spoken and if census forms are not in the correct one it will be impossible to fill in. For example, India has 22 official languages (although it is estimated that approximately 800 different languages are actually used throughout the country). India required 1.7 million enumerators in 1991. To make forms in many languages is costly.

5. War zones are unable to be included due to the danger. Wars also create **refugees** who could be counted more than once or not at all. **War fatalities** may already have been counted.

6. **Migration** has a large effect on population, and in times of extreme hardship vast numbers of people can move into or out of troubled areas, making previous counts largely invalid, e.g. rural–urban migration sees 300 families migrate into Mumbai (Bombay) every day! **Nomadic people**, like the Bedouin, can be omitted from census counts or counted repeatedly. **Illegal immigrants** (such as illegal Bangladeshi immigrants in India, for example) may deliberately avoid census counts.

7. Some people are simply **distrustful of governments** and don't see why they should provide this information. They may not return accurate data.

8. **Under registration** can occur for **social** and **religious** reasons; for example, women may not always be included in census returns in certain Islamic countries.

The future?

'Our population is changing rapidly, and we will always need good population statistics. There are different ways we could take a census in future.'

Peter Benton

Beyond 2011 Programme Director

In 2013, the Office for National Statistics (ONS) conducted a survey to find out how future population statistics could be gathered. Population statistics are vital to allow the government to plan for the future. Population continually changes so there will always be a need to gather information. However, how these statistics should be gathered is under consideration. With the advances in technology, the following methods are under consideration for the future collection of data.

- A census once a decade, like the one conducted in 2011, but primarily online.
- A census using existing government data and compulsory annual surveys.

Both methods would produce the necessary data for government planning. The annual surveys and existing data would provide relevant, current data each year, whilst the full 10-year, online census would give more detailed statistics once a decade.

Make the link

You may have learned about statistics in Maths.

Make the link

Like census takers, you will collect and process information for your Assignment.

Population structure

The information gathered from a census shows how many men, women, young, middle-aged and elderly people there are in the country. This can be used to produce graphs called **population pyramids** which show the population's structure. Different countries have different structures and subsequently they have different shapes of population pyramid. An example is shown below:

Population pyramid for a developing country

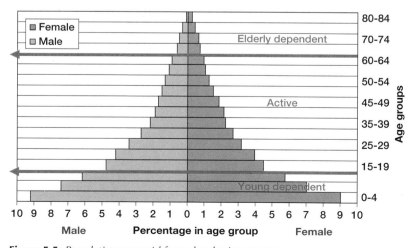

Figure 5.5: *Population pyramid for a developing country*

The population is divided into:

- male and female
- age groups of five-year bands
  - 0–4,
  - 5–9,
  - 10–14,
  - 15–19 ... 80+
- 3 broad groups
  - children 0–14 – too young to work – YOUNG DEPENDENT
  - active 15–64 – work and support others – ACTIVE
  - aged 65+ – senior citizens, retired – ELDERLY DEPENDENT

Population pyramids can tell us important facts about the structure of the population of that country. They can also tell us about the **birth rate, death rate** and **life expectancy**.

- Birth rate: the number of births per 1000 people per year.
- Death rate: the number of deaths per 1000 people per year.
- Life expectancy: the average number of years that individuals are expected to live depending on where and when they are born and spend their lives.

⠿ Make the link

You will learn more about how 'indicators' like birth and death rates can help us to determine the development of a country in the Development and health chapter.

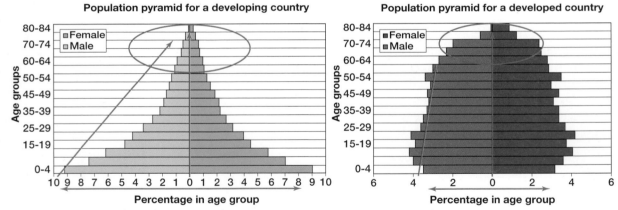

Figure 5.6: *Structure of population pyramids*

1. **Birth rate** – this is shown by the **width of the base**. The wider the base of the pyramid, the higher the birth rate.
2. **Death rate** – this is shown by the **sides of the pyramid**. The more sloping it is, the higher the death rate as there are less people in each successive age group.
3. **Life expectancy** – this is shown by **how high the pyramid reaches** and how many people have survived into old age.

Population structure of a developing country (ELDC)

Population pyramid for a developing country

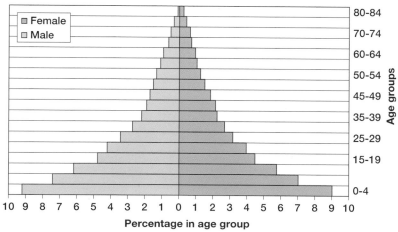

Figure 5.7: *Population pyramid of a developing country*

Figure 5.8: *Developing countries usually have a high proportion of young dependents*

An economically less developed country (ELDC) has a **triangular** shaped pyramid because:

Describe	Explain
It has a **wide base** indicating a high proportion of young dependents.	This is due to **high birth rates** caused by: • limited access to family planning • high infant mortality rate • need for workers in agriculture • religious beliefs • children viewed as economic assets • large family traditions.
The pyramid **narrows quickly** indicating a decreasing working population.	This is due to **high death rates** caused by: • high levels of disease • famine • lack of clean water and sanitation • lack of health care/poor access • lack of education • war.
The pyramid **tapers rapidly** towards the top indicating a low percentage of elderly dependents.	This is due to **high, but falling, death rates** and **low life expectancy** caused by: • poor geriatric health care • inadequate welfare systems • overcrowding • malnutrition.

Population structure of a developed country (EMDC)

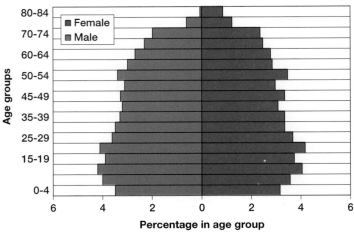

Population pyramid for a developed country

Figure 5.9: *Population pyramid of a developed country*

An economically more developed country (EMDC) has a **barrel** shaped pyramid because:

Describe	Explain
It has a **narrow base** indicating a low proportion of young dependents.	This is due to **low/falling birth rates** caused by: • increased family planning available • increased availability of contraception • lower infant mortality rate • increased mechanisation reduces need for workers • increased standard of living • changing status of women • materialistic lifestyles.
The pyramid has a fairly **uniform structure** with **filled out sides** indicating a high percentage of working population.	This is due to **low death rates** caused by: • accessible and widespread health care (e.g. smallpox vaccine, penicillin) • improved hygiene (drinking water boiled) • improved sanitation • improved food production and storage.
The pyramid has a **reasonably wide top** indicating a large proportion of elderly dependents.	This is due to **declining death rates** and **increased life expectancy** caused by: • advances in geriatric health and community care • state and private pensions • rising living standards • specialised housing • improved diets.

Figure 5.10: *EMDCs usually have a high proportion of elderly dependents*

Consequences of population structure

Aging populations

Most developed countries (e.g. the UK) are in Stages 4 or 5 of the Demographic Transition Model where their birth rates and death rates are low and their populations are rising slowly, if at all. The death rate is so low that there are many people living into old age. While less money needs to be spent on education (as there are fewer children), this structure causes many problems.

Effects

1. Need for more care services (e.g. day care, meals on wheels).
2. Need for more sheltered housing and old people's homes.
3. Cost of health care rises.
4. Increased pressure on working population as costs are paid for by taxes.
5. Fewer people of working age to pay taxes.
6. Fewer people for the armed forces.
7. Fewer potential parents for the next generation.
8. Rise in retirement/pension age.

Solutions

1. More paternity leave to encourage people to have children.
2. More maternity benefits (incentives).
3. Raise retirement age = increase number of taxpayers and reduce pensions.
4. Encourage more women to work = increase number of taxpayers.
5. Provision of crèches in the workplace.
6. Allow more immigrants into the country = increase number of taxpayers.
7. Encourage more people to take out private pension schemes, to reduce the cost of providing public pensions.

<aside>
Make the link

The Demographic Transition Model is covered in detail as part of the National 5 Geography course.
</aside>

<aside>
Hint

Questions are often linked, asking for effects and strategies.
</aside>

Figure 5.11: *One solution to the problems of an ageing population is to encourage more women into work, increasing the number of taxpayers*

Make the link

If you take Modern Studies, you will learn more about the position of women in society, both in the UK and in other countries.

Make the link

You will learn more about issues in developing countries such as these in the Rural, Urban and Development and health chapters.

Figure 5.12: *One strategy used to tackle the problems of a population with high birth rates is the greater education of girls*

Youthful populations

Most developing countries (e.g. Gambia) are in Stages 2 or 3 of the Demographic Transition Model where their birth rates are much higher than their death rates and their populations are rising rapidly. There are many children because the birth rate is high. Often, half of the population is under 15 years of age. This structure also causes many problems.

Effects

1. Increased pressure on working population to provide for more than half of the population.
2. Need to spend a lot of money on hospitals, doctors and nurses to provide the medical care needed by children.
3. Expense of providing schools and teachers.
4. Pressure on farmers to grow enough food.
5. Land farmed intensively, making soil poorer.
6. More and more trees are being cut down to create farmland = desertification.
7. People become poor and hungry due to stretched resources, and many move to cities in search of work.
8. Not enough housing for everyone so people build their own makeshift shacks which lack basic amenities, e.g. toilets and water supply.
9. Insufficient jobs for everyone so unemployment is high and crime-rates rise. Rise in informal sector employment.
10. Traffic congestion worsens as city populations increase.
11. Schools and hospitals are overcrowded and not everyone has access to them.

Solutions

1. Laws limiting family size, e.g. China's One Child Policy (1979). In 2016 this was changed to a two child policy.
2. More education on reducing number of births, e.g. family planning clinics.
3. Greater education of females (evidence of lower birth rates).
4. More opportunities for abortions and sterilisations.
5. Incentives given to limit family size, e.g. free health care, preferential housing.
6. High-yielding crops, fertilisers, pesticides and irrigation used to improve farming and crop yields.

Migration

Migration is the movement of people from one place to another. This can be from one area of a country to another, or it can be international – from one country to another. Migration can be voluntary, when people move because they want to, or it can be forced, when people have no choice but to leave their homes.

Voluntary migration is when people choose to move. This is usually because of **pull factors.** Some areas **attract** people and **pull** people towards them.

Forced migration is when people have no choice but to move to a better area. They can be **pushed** out of an area because of religious or political problems or natural disasters. These people are called **refugees.**

Push and **pull** factors can be used to describe the reasons why people move.

Make the link

If you take History or studied it at National 5, you may have learned about migration into and out of Scotland in the 19th and 20th centuries.

Push factors	Pull factors
• Natural disasters (e.g. drought)	• More job opportunities
• Infertile soils and crop failure	• Higher wages
• Lack of services	• Better housing and basic services (water, sewerage, electricity)
• Low wage rate	• More educational opportunities
• High unemployment rate	• Food available in shops
• Poor health services	• More opportunities in a big city
• Lack of educational opportunities	• Plenty of services available (hospitals, schools)
• Poor food supply (need to grow own)	• More entertainment available
• War, famine, disease	

Forced migration

Figure 5.13: *Syrian refugees at the Turkish border*

It is estimated that there were **22.5 million refugees** worldwide at the end of 2017. A refugee is a person who has been forced to leave their home because of war, natural disasters or persecution due to race, religion, nationality or political opinion. An **asylum seeker** is a person fleeing persecution who is waiting to see if they are to be granted refugee status.

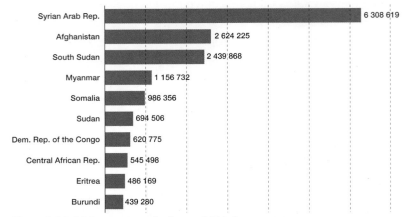

Figure 5.14: *Major sources of refugees 2013, by country*

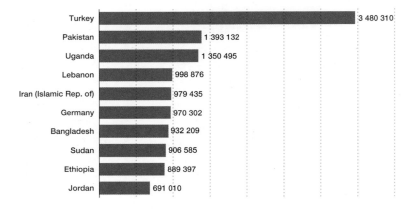

Number of admitted refugees

Figure 5.15: *Major refugee hosting countries*

Figures 5.14 and 5.15 show the major countries producing and receiving refugees. It can be seen that developing countries host more than 80% of the world's refugees. People fleeing conflict or persecution often go to the nearest country for refuge but they may not necessarily want to stay there permanently.

Turkey, with 3.48 million, ranks first for taking refugees. Most Syrian refugees are in Turkey and neighbouring Lebanon, Jordan and Iraq. The vast majority of Afghan refugees are in Pakistan and Iran (see Figure 5.17).

Forced migration case study: Afghanistan

Make the link
You may have studied terrorism in Modern Studies.

Since 1978, Afghanistan has experienced three and a half decades of huge political instability and conflict **due to war (push factor)**. From 1978 to 1997, civil war left Afghanistan politically and economically unstable. In 1997, the radical Islamic group the **Taliban** seized control of the capital city Kabul and declared themselves the new political and religious leaders in Afghanistan. Following the 9/11 terrorist attacks in the USA in 2001, the Taliban refused to deport Osama Bin Laden. This resulted in a military campaign led by the USA and the removal of the Taliban from government in December 2001. Since 2001, Afghanistan has been attempting a process of **reconstruction,** which recognises all Afghans as equal. However, the Taliban has continued to be an insurgent force within Afghanistan and violence continues to this day. This has forced millions of people to flee Afghanistan and become refugees. The majority of refugees make their way to Pakistan. Other surrounding countries of Iran, Turkmenistan, Uzbekistan and Kyrgyzstan also take refugees from Afghanistan.

The case is similar in Syria, where civil war is creating large numbers of refugees, many of whom come to Europe. Many also apply for asylum. Causes of forced migration from Syria include loss of services during the ongoing conflict, fear of persecution and lack of employment due to damage to infrastructure. All these factors result in the creation of ghost towns.

Figure 5.16: *An Afghan girl carries her belongings at a refugee camp in Pakistan*

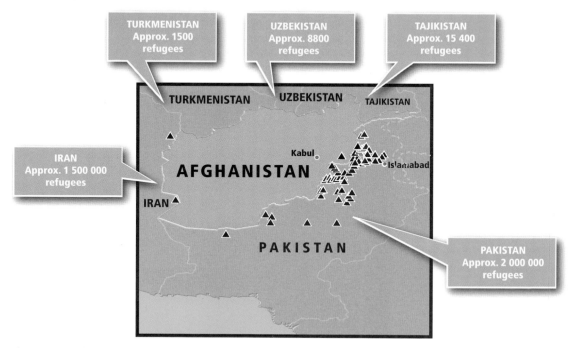

Figure 5.17: *Location of refugee camps*

This forced movement of people from Afghanistan to surrounding countries has consequences for both the donor country (Afghanistan) and the receiving countries (e.g. Pakistan).

Hint

Revise your case studies carefully. Including place names, figures and statistics in your exam answers can help you gain marks.

Consequences for Afghanistan (Donor)

- There are fewer people to grow and tend to crops, resulting in food shortages.
- There are not enough people left with the skills to provide basic services such as health care and education, or work in government. There is also a lack of funding for these areas.
- The same is true of law enforcement, which also has problems with corruption.
- Lack of funding and workers has also affected transportation – the roads have deteriorated and other forms of transport have become difficult and dangerous.
- There is a lack of economic opportunities, basic services, sufficient food and clean water. This is especially true in rural areas and affects families that never left, or have since returned to Afghanistan.

Figure 5.18: *Afghan refugees*

Consequences for Pakistan (Receiver)

- Pakistan currently has the world's second largest refugee population.
- Afghan refugees in Pakistan's western borderlands pose difficult problems for the government as they need to be provided with food, clothing and shelter, which is a financial burden on the government.
- Ongoing violence between Afghans and Pakistanis; Pakistanis have been injured and killed in conflicts over land use and access to resources such as water.

The United Nations refugee agency provides up-to-date figures and shows that more than 6.2 million refugees have voluntarily returned to Afghanistan in the last 10 years, of whom more than 4.6 million were assisted to do so by the UN High Commissioner for Refugees (UNHCR). Nonetheless, some 2.7 million Afghans continue to live in exile in neighbouring countries.

> ### 🔍 Hint
> Watching the news can be a good source of up-to-date information e.g. on Syrian refugees.

Figure 5.19: *The UNHCR headquarters in Geneva, Switzerland*

> ### ❓ Did you know?
> By June 2019, about 6.7 million Syrians were refugees, and another 6.2 million people were displaced within Syria.

Asylum seekers

Some migrants fear for their lives and decide not to return to their own country. Syria's civil war is resulting in large numbers of migrants moving to Europe along with people from Afghanistan and Iraq. Europe is easier to reach overland or by boat, especially for those fleeing conflicts in the Middle East or Africa, and people traffickers already have well-established smuggling routes to Europe. Figure 5.20 shows the hazardous journey faced by Afghans trying to seek asylum in Europe. In 2017, more than 2275 migrants died crossing the Mediterranean and in one single incident in 2012, 300 migrants drowned off the coast of Malta. In 2018, there were around 2.5 million registered refugees from Afghanistan.

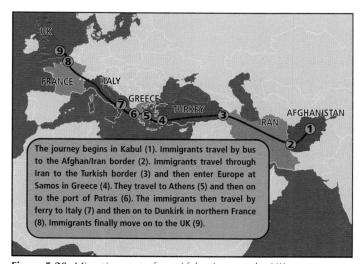

The journey begins in Kabul (1). Immigrants travel by bus to the Afghan/Iran border (2). Immigrants travel through Iran to the Turkish border (3) and then enter Europe at Samos in Greece (4). They travel to Athens (5) and then on to the port of Patras (6). The immigrants then travel by ferry to Italy (7) and then on to Dunkirk in northern France (8). Immigrants finally move on to the UK (9).

Figure 5.20: *Migration route from Afghanistan to the UK*

The future?

At the end of 2014, NATO officially ended combat operations in Afghanistan although in 2019, the United States still had around 14 000 troops in the region. Foreign aid to the area is also set to decline, which could mean that the Afghan economy could face an economic crisis. These factors might mean that many Afghans will still seek a better life elsewhere.

Voluntary migration case study: Poland to the UK

On 1 April 2004, Poland was one of 10 countries admitted to the EU. Of these countries, eight were eastern European – Estonia, Latvia, Lithuania, **Poland**, Czech Republic, Slovakia, Slovenia and Hungary (the other two were Malta and Cyprus).

Under the law, any resident of the EU is free to move to any other EU country. As unemployment was higher and the standard of living was lower in the eight eastern European countries than in the UK, it was predicted that a lot of people would migrate from these countries to the UK.

The UK government expected that 15 000 migrants would come from the eastern European countries to the UK for employment. In the event, by July 2006, 447 000 people from eastern Europe had applied to work in the UK, 62% (264 555) of which were from Poland, increasing to 370 000 by the end of 2006. The Polish embassy estimated the number of Poles in the UK to be between 500 000 and 600 000, meaning Polish people would be the third largest ethnic minority in the UK.

Figure 5.21: *Migration map: Poland to the UK*

Figure 5.22: *The number of Polish grocery shops in the UK has increased to meet the needs of the growing Polish population.*

Causes of voluntary migration

Push factors	Pull factors
Average unemployment in Poland of 18.5% in 2005.	The unemployment rate in the UK was 5.1%.
Rural unemployment of over 40% in some areas.	The GDP was $30 900 per person in the UK, in Poland it was $12 700.
Approximately 40% of young people were unemployed.	The UK had a high demand for skilled and semi-skilled labour: vacancies in the UK for Oct–Dec 2007 were 607 900.
Average wage much lower than in the UK.	The UK was one of only three countries that did not place a limit on the numbers of immigrants allowed to come from the eastern European countries.

The migrants:

- Skilled and semi-skilled industrial workers and tradespeople.
- On average earned around £150 per month in Poland and £6–7 per hour in the UK.
- Most were employed within factories, warehouses, farms or as cleaners.
- Most intended to stay for less than a year.
- Often had young families.

Make the link

In Modern Studies, you may have looked at unemployment in the UK and in other countries.

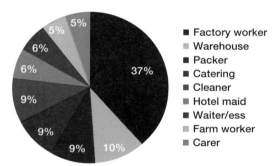

- Factory worker
- Warehouse
- Packer
- Catering
- Cleaner
- Hotel maid
- Waiter/ess
- Farm worker
- Carer

Figure 5.23: *Top 10 migrant jobs*

Consequences of voluntary migration

This migration of Poles to the UK has been a benefit and a problem to both Poland (the donor) and the UK (the host). The table overleaf displays some of these benefits and problems for Poland and for the UK.

Poland

Benefits	Problems
• Reduced pressure on employment and resources, e.g. food, water.	• Poland is losing high numbers of their skilled, ambitious and educated active population.
• Polish migrants develop new skills which will benefit the Polish economy if they return.	• Gaps in their workforce are having to be filled by the elderly, reducing outputs and affecting the economy.
• Polish migrants earn 'hard currency' – the UK pound – which can be sent back to Poland and boost the economy.	• Mostly young Polish men have migrated, which reduces the numbers available to defend Poland.
• Mostly young Polish men have migrated, which has reduced the birth rate.	• As mostly males have migrated, families have been divided, causing domestic stress and tension.

UK

Benefits	Problems
• Polish migrants are prepared to do menial, unskilled, low-paid jobs in factories, catering, cleaning, labouring, which many UK citizens do not want to do.	• Polish workers may be exploited by some employers.
• Polish workers are prepared to work long, unsociable hours.	• Increase in population in some areas will result in higher rents/prices as demand for housing rises.
• Polish National Insurance contributions have helped the UK cope with its aging population.	• Increased pressure on education services caused by the need to provide schooling for the children of immigrants.
• Polish people have enriched UK culture – customs, festivals, food, etc.	• Increased demand on the NHS.
	• In areas that have not seen much immigration in the past, there may be tensions.

Figure 5.24: *Many Polish delis have opened across the UK*

The current condition

The number of people migrating to the UK from Poland has reduced as the cost of living in the UK has increased and conditions in Poland have improved; roughly half of the 'original' eastern European migrants have already returned home to their countries.

🔍 **Hint**

When you are revising your case studies, try to find out what the current situation is so that you can write a really up-to-date answer in the exam.

Summary

In this chapter you have learned:

- Ways in which governments collect data on population
- Why countries conduct a census
- Problems of conducting a census in developed and developing countries
- Collecting population data in the future
- Differences between the population structure of a developed and developing country
- The consequences of a changing population structure for developed and developing countries
- Differences between forced and voluntary migration.

You should have developed skills and be able to:

- Interpret numerical and statistical data from a census
- Analyse and evaluate data from a census
- Analyse population structure from a population pyramid
- Interpret the differences in population structure for a developing and developed country from a population pyramid
- Interpret migration routes from a map
- Analyse migration issues from a variety of sources.

End of chapter questions and activities

Quick questions

1. Look at Figure 5.25. Discuss the reasons why conducting a census could be difficult and why population data taken from census records could be unreliable.

a. Bomb damage, Syria

b. Mountain village, Nepal

c. Rainforest tribe dwelling, Amazon

d. Chad: education breakdown

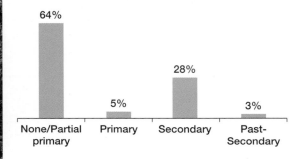

Figure 5.25: *Selected world areas*

2. 'I don't see why the UK needs a census.'

   Explain the ways in which the UK government benefits from having census data.

3. 'I was OK at the beginning, but as I started living here life got difficult. This is not the life I dreamed of.'

   Look at the quote above from a Polish migrant. Explain the problems facing Polish immigrants coming to the UK.

4. Explain the differences between forced and voluntary migration.

5. Explain the push and pull factors that cause migration.

Exam-style questions

1. With reference to an international migration you have studied, **explain** the impact on both the donor and receiving country.

    8 marks

2. Study Figure 5.26.

    Discuss the possible consequences of the 2050 population structure for the future economy of Malawi and the welfare of its citizens.

    9 marks

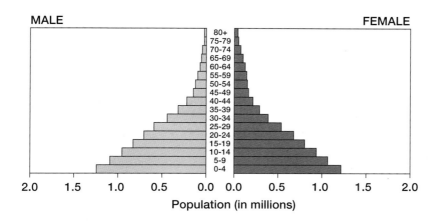

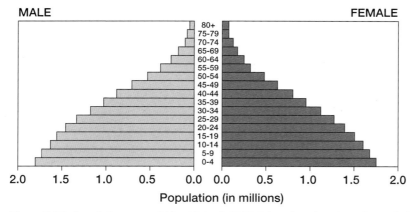

Figure 5.26: *Population pyramid for Malawi, 2010 (above) and projected population pyramid for Malawi, 2050 (below)*

Activity 1: Presentation

In groups, create a presentation on the problems of conducting a census in various countries around the world and here in the UK. Your presentation should be roughly 10 minutes in length. You can use PowerPoint to help if you wish. Be prepared to answer questions from other pupils at the end of your presentation.

Activity 2: Poster

THE NEW BRITONS

Migration brings benefits and problems to the UK

Since joining the EU, hundreds of thousands of Polish people have come to the United Kingdom. Critics say they deprive Britons of jobs and houses. Economists say they add £300m a year to the economy.

Look at the newspaper article 'Migration brings benefits and problems to the UK'.

Divide into groups – each group chooses to discuss either the benefits or problems migration brings to the UK. Divide a sheet of poster paper into two halves. Head one side 'Benefits' and the other 'Problems'. Note your findings on the poster paper. Exchange your findings with the rest of the class. Add to your poster extra information shared from the class.

Activity 3: Debate

'An ageing population will cause more problems for a government than a youthful population.'

The class will be split into two groups, one will argue in support of the above statement and one will argue against.

Activity 4: Evaluation

With a partner, evaluate the impact of either a forced migration (e.g. Afghanistan) or a voluntary migration (e.g. Polish to UK) on the donor and receiving country.

Copy the table and complete to summarise your findings.

Economic, political, social and environmental impact of migration	Donor country	Receiving country
Economic impacts		
Political impacts		
Social impacts		
Environmental impacts		

In conclusion, write a short speech putting forward your opinion on which country (donor or receiver) is impacted most by the migration. Be prepared to justify your opinion to the rest of the class.

Learning Checklist

You have now completed the Population chapter. Complete a self-evaluation sheet to assess what you have understood. Use traffic lights to help you make up a revision plan to help you improve in the areas you have identified as amber or red.

- Demonstrate an understanding of the methods used to collect population data.

- Explain the reasons why governments collect population data.

- Discuss the problems of conducting a census in both a developed and developing country.

- Explain the structure of a population pyramid using terms such as dependent and active population as well as birth rate, death rate and life expectancy.

- Using population pyramids, be able to compare the population structure of a developing and developed country.

- Using a named example, explain the consequences of population change in a developed country (aging population).

- Using a named example, explain the consequences of population change in a developing country (youthful population).

- Suggest ways to deal with the effects of population change in both a developed and developing country.

- Understand the terms migration, voluntary migration, forced migration, push and pull factors, refugee and asylum seeker.

- Discuss the push and pull factors that influence migration.

- Outline the differences between voluntary and forced migration.

- Referring to a case study, discuss the consequences of a forced migration on the donor and receiving country.

- Understand the differences between an asylum seeker and a refugee.

- Referring to a case study, explain the push and pull factors involved in the voluntary migration chosen.

- Discuss the benefits and problems to both the donor and receiving countries.

Glossary

Ageing population: a country has an ageing population when there is a high proportion of elderly people.

Birth rate: the number of births per 1000 people in one year.

Census: a counting of people by the government every 10 years.

Contraception: the use of birth control to stop pregnancy.

Death rate: the number of deaths per 1000 people in one year.

Demographic Transition Model: a model showing the stages of growth of a country's population.

Density: the number of people per square kilometre.

Dependency ratio: the ratio between those of working age and those of non-working age.

Depopulation: the decline of population in an area.

Distribution: the spread of people over an area.

Economic migrant: a person leaving their own country to seek better economic opportunities and so settle temporarily in another country.

Emigrant: a person who leaves an area to live elsewhere.

Enumerator: a person employed in taking a census of the population.

Family planning: the use of contraception to control the size of your family.

Immigrant: a person who moves into an area from elsewhere.

Infant mortality: the number of babies dying before their first birthday per 1000 live births.

Life expectancy: the average number of years a person might be expected to live.

Literacy rate: the proportion of a population able to read and write.

Migrant: a person who moves from one place to another to live.

Migration: movement of people.

Natural increase or decrease: the difference between the birth rate and the death rate.

Population change: Births – Deaths + In-migration – Out-migration = Population change.

Population density: number of people per square kilometre.

Population pyramid: a graph that shows the age and sex structure of a place.

Push–pull factors: push factors push people out of an area and pull factors attract people to an area.

Refugees: people forced to move from where they live to another area.

Voluntary migration: the movement of people from one area to another area through choice.

6 Rural

Within the context of Rural you should know and understand:

- The impact and management of rural land degradation related to a rainforest or semi-arid area
- Rural land use conflicts and their management related to either a glaciated or coastal landscape.

You also need to develop the following skills:

- Identifying areas of degradation from maps
- Interpreting diagrams
- Extracting information from photographs.

Make the link

You will learn about soil in detail in the Biosphere chapter.

Hint

You may have studied the causes of deforestation in the rainforest in National Geography 5. These will not be examined at Higher level, but you will need to be familiar with them to discuss their impacts and management.

Soil erosion

Soil erosion is the loss of soil from the land by the action of wind and water. This is usually a slow, natural process that allows time for new soil to be made to replace it. However, humans can speed up the process by carrying out the developments shown in the table below.

Causes of rural land degradation	
Africa, north of the Equator	**The Amazon Basin**
Deforestation	Deforestation
Over-cultivation	Cattle ranching
Over-grazing	Mining
Population increase	Hydroelectric power (HEP) schemes

Figure 6.1: *Grazing land in the Sahel region*

Figure 6.2: *Deforestation in the Amazon Basin*

Erosion reduces the fertility of the soil and this causes a decrease in crop production. On average, it takes 500 to 1000 years to form one inch of topsoil naturally. In Africa, 65% of agricultural land has been lost to degradation while in South America, this percentage is 45% due to the developments shown previously.

> **? Did you know?**
>
> Most streams and rivers in the Sahara are only seasonal. The main exception is the Nile River. It crosses the Sahara and empties into the Mediterranean Sea.

Africa: semi-arid area – Sahel case study

Background

The Sahel is an area of land lying to the south of the Sahara Desert which suffers from increasing land degradation and desertification. It stretches from Mauritania to Ethiopia (see Figure 6.3). The Sahel is roughly 5400 km long, covers an area of about 3 million km² and receives between 200 mm and 600 mm of rain annually. The vegetation is mainly grass and shrub land and the people are traditional semi-nomadic herders (Figure 6.4).

> **? Did you know?**
>
> The word 'Sahel' means 'shore' in Arabic.

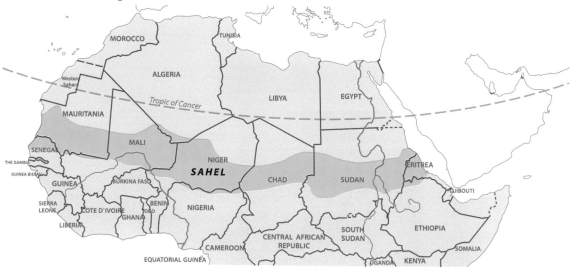

Figure 6.3: *Location of Sahel region*

? Did you know?

Around 32 million people suffer from food shortage in the Sahel.

Make the link

There is more information about migration in the Population chapter, and you will learn more about shanty towns in the Urban chapter.

Impact on people and the environment

One of the consequences of land degradation in the Sahel is crop failure, which has led to people being **under-nourished** and **deaths from starvation**.

These famine conditions have badly affected countries such as Mauritania, whose agricultural zone has shrunk to a 200-km-wide strip. When people are under-nourished, they are susceptible to diseases such as kwashiorkor. **Disease and illness become widespread** so people **cannot work** so have **no money to buy food**. They **become weaker** and a cycle of poverty develops (Figure 6.5). The people in rural areas are **forced to leave the countryside** and move to the cities in search of food and employment. Many of these people cannot find anywhere to live or any employment and **end up living in shanty towns** on the edge of the city.

Figure 6.4: *Traditional semi-nomadic herders*

The **traditional life of the nomads is under threat** as they cannot find food and water for their animals. Many are forced to settle in villages or at oases. This in turn puts pressure on the surrounding land, leading to **over-cultivation** and soil erosion. This also leads to the spread of the Sahara Desert. In the last 30 years, the proportion of Mauritania's people living in the capital Nouakchott rose from 9% to 41%, while the proportion of nomads fell from 73% to 7%. Many people are forced to migrate, leaving their homes and seeking food and shelter in neighbouring countries. This can lead to **conflict with the resident populations**. International aid is often necessary to ensure the people's survival but this can lead to an **over-dependence on aid**.

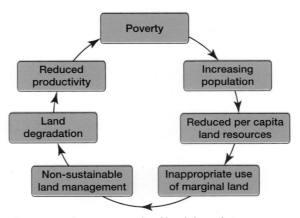

Figure 6.5: *The vicious cycle of land degradation*

Management strategies

To try to reduce the impact of land degradation, some of the following methods of soil conservation can be used.

1. **Movable fencing** allows farmers to restrict grazing animals to specific areas of land, which enables remaining land to recover. This allows farmers to move animals between fenced areas, reducing the dangers of over-grazing and trampling of soil and allowing soil and land to recover between grazing sessions.

Figure 6.6: *Movable fencing*

🔍 Hint

Do not list the strategies. Answers need to be explanations, so give as much detail as possible.

2. **Contour ploughing** can be used to prevent soil being washed downhill. Contour ploughing is the practice of ploughing across a slope following its contour lines. The rows formed slow down water run-off during rainstorms, preventing soil erosion and allowing the water time to settle into the soil.

3. A **terrace** is a levelled section of a hill, designed to slow or prevent the rapid surface run-off of irrigation water.

4. Similarly, **'magic stones'** are lines of stones laid along the contours of gently sloping farmland to catch rainwater and reduce soil erosion. 'Magic stones' allow the water to seep into the soil rather than run off the land. This prevents soil being washed away and can double the yield of crops such as groundnuts.

Figure 6.7: *Contour ploughing*

Figure 6.8: *Terracing*

5. Trees are planted in rows as **shelter belts** to stop the wind drying out the ground and blowing the soil away. An example of this is the Great Green Wall in Africa. Also, the roots bind the soil together. Trees can be grown for shade to allow crops to be grown in between the rows.

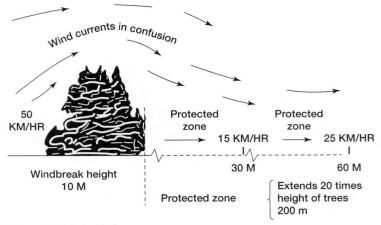

Wind currents in confusion

50 KM/HR

Windbreak height 10 M

Protected zone → 15 KM/HR →

Protected zone 25 KM/HR

30 M 60 M

Protected zone { Extends 20 times height of trees 200 m

Figure 6.9: *Shelter belts*

6. **Strip cultivation** is also used. This is when small crops are grown between tall crops for shelter.

Figure 6.10: *Strip cultivation*

7. In drier areas, **irrigation** is used (this is the artificial watering of crops from stored water). This keeps the soil moist, allowing crops to grow and preventing soil being blown away.

Figure 6.11: *Irrigation*

8. **Planting pits** or **zai**. Farmers dig a grid of planting pits. Run-off water is trapped in the hole, then manure or compost is added to the hole. Termites are attracted and they dig galleries that make the infiltration of rainwater, run-off and the retention of moisture easier.

9. **Stone bunds** form a barrier that slows down water run-off, allowing rainwater to seep into the soil and spread more evenly over the land. This slowing down of water run-off helps with building up a layer of fine soil and manure particles, rich in nutrients.

Figure 6.12: *Planting pits*

Success of methods of soil conservation

Some of these methods have been successful. Managed grazing areas are successful if fencing is available and affordable. However, agreement needs to be reached by the herders in the area, which is not always possible. For example, in the settlement of Korr in northern Kenya, herders harvest woody thorn materials mainly for construction of livestock enclosures to secure their livestock, but this process has led to destruction of whole woodlands within the vicinity of settlement areas and along the routes of livestock movement, leaving some herders with no food supply for their animals.

Figure 6.13: *Stone bunds*

Figure 6.14: *African cattle in an enclosure*

Contour farming can reduce soil erosion by as much as 50% on moderate slopes but for slopes steeper than 10%, other measures need to be combined with contour farming to increase its effectiveness. However, if contour lines are not correctly established, then they can in fact increase the risk of erosion. Terraces are successful as they do slow down the flow of irrigation water, so more water is absorbed, keeping the soil moist.

'Magic stones' have little cost and the trapped soil can be raked over the land, making use of the trapped minerals. Farmers can help each other. In Mali and Burkina Faso, this method has been successful and some crop yields have increased by as much as 50%. Shelter belts around farmland help reduce the effects of sandstorms, wind erosion, shifting sand and droughts. They reduce the effects of temperature, wind speed, soil water loss and create more favourable conditions for crop production. The produce from the trees can also be sold, and the wood from the trees can be used for fuel and building materials when they are eventually cut down.

Irrigation is successful if water is available to be stored. However, overuse of irrigation water can make the soil saline, thus reducing the fertility of the soil and resulting in the need to use fertilisers, which poor farmers cannot afford. Some irrigation methods are too expensive for farmers to use; for example, drip irrigation and fertilisers.

Zai is a simple technique that needs no equipment other than what is available on the land. Zai are usually constructed on abandoned or unused ground so crop yields resulting from this practice bring a benefit of 100%. Stone bunds are successful as they retain water and silt, building up new soil.

Brazil: Amazon rainforest case study

Background

The Amazon rainforest is a tropical rainforest located along the Equator in South America (Figure 6.15). It is the biggest rainforest in the world. Tropical rainforests are usually warm and wet with temperatures averaging around 27°C and rainfall greater than 2000 mm per year. There are no seasons in the rainforest – they are hot and humid all year. The high rainfall and year-round high temperatures are ideal conditions for vegetation growth. The wide range of plants also encourages a huge variety of insects, birds and animals to live here. It is estimated that about half of all the Earth's plant and animal life lives in rainforests. Rainforests like the Amazon are extremely important in the ecology of the Earth. The plants of the rainforest generate much of the Earth's oxygen; without oxygen all living things on Earth would die.

Impact on people and the environment

One of the consequences of land degradation in the Amazon rainforest is the destruction of the way of life of the indigenous people, e.g. clashes between the Yanomami rainforest tribe and incomers to the rainforest who are there to exploit it for money. These incomers often

Hint

Naming tribes and including a case study is often worth one or two marks in case study-based questions in addition to each factor you include in your answer.

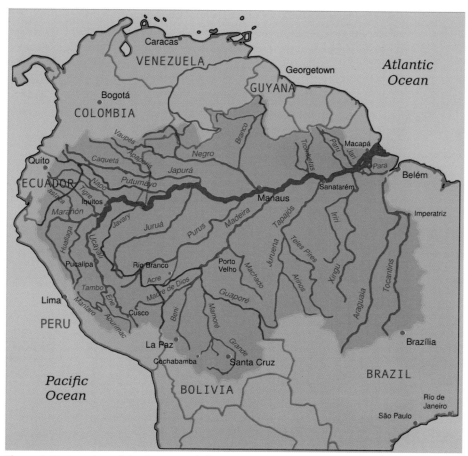

Figure 6.15: *Location of the Amazon rainforest*

bring with them 'western diseases', like the common cold, which the tribal people have no immunity or resistance to as they have lived in isolation for so long. These tribes are generally subsistence farmers who rely on the land to feed themselves and their families.

Land degradation caused by the removal of the forest results in nutrients being washed from the soil as the canopy of trees is no longer there to protect the soil from the heavy rains (Figure 6.18). The loss of nutrients from the soil means that over time the soil will become less and less fertile, leading to a reduction of fallow period and therefore reduced crop yields for food. These tribal people may eventually be forced to leave their rainforest homes and move to reservations created by the Brazilian government. Some may choose to leave the rainforest completely and migrate to the city in search of work to support their families. The loss of their traditional way of life often results in increased alcoholism among the indigenous people.

Land degradation can also result in a loss of wildlife habitats and overall a loss of biodiversity – some species may even be threatened

Figure 6.16: *Yanomami natives have been attacked and killed by those illegally mining for gold in the past*

⚆ Make the link

You have already looked at the relationship between rain and soil in both the Hydrosphere and Biosphere chapters.

Figure 6.17: *Deforestation can result in a loss of wildlife habitats*

with extinction. There are many plant species in the rainforest that are yet to be discovered. These may hold the key to creating cures for illnesses like cancer. If deforestation continues, potentially useful medicinal drugs may be lost forever. The rainforest is said to be the 'lungs of the world'. The plants there take in the carbon dioxide we breathe out and replace it with the oxygen we breathe in. Without these plants to remove the carbon dioxide from the atmosphere, there may be an impact on our global climate (greenhouse effect).

Make the link

The greenhouse effect is covered in the Global climate change chapter.

If you take Biology, you will have learned more about the importance of biodiversity.

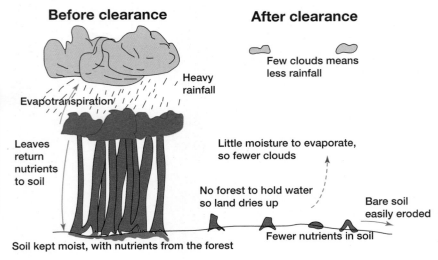

Figure 6.18: *Deforestation*

Summary

	The effects of land degradation
Wildlife	Loss of habitat, extinction of species
Climate	Increase in CO_2 (greenhouse effect); rise in global temperatures; rise in sea levels
Local people	Loss of habitat, source of food supply, way of life
Soil	Loss of fertility; soils can be washed away by the heavy rains

Methods of soil conservation/reduction of land degradation

Charles, Prince of Wales, has been an avid campaigner to preserve the world's rainforests. He set up the Prince's Rainforest Project in 2007 after becoming concerned about global climate change and how the destruction of the world's rainforests is contributing to this. If deforestation can be minimised or stopped altogether, so too can land degradation in these areas.

Figure 6.19: *Charles, Prince of Wales, on a visit to the Amazon rainforest*

Make the link

You may have learned about pressure groups that focus on the environment in Modern Studies.

'My Rainforests Project has been working to try and find ways to ensure the trees become more valuable alive than dead, rather than the other way round, so that there is no incentive to cut them down. Following exhaustive consultation with the private, public and N.G.O. sectors over the past eighteen months, and working with experts in the various relevant fields, we have been attempting to develop a solution.'

Charles, Prince of Wales (5th May 2009, London)

Methods include:

1. **Agroforestry** is the growing of both trees and crops on the same piece of land. This is designed to provide tree and other crop products, whilst protecting and conserving the soil. It allows the production of diverse crops, which benefits both the land and people. Examples of crops are wood products, Brazil nuts, cacao and guarana. Agroforestry increases infiltration, which reduces rainspalsh and sheetwash of soil.

Figure 6.20: *Agroforestry*

2. **Afforestation** by conservation groups, both national and international, aims to conserve soils by reforestation and the protection of existing forests. For example, the Amazon Region Protected Areas (ARPA), created in 2002 by the Brazilian government in partnership with the World Wildlife Fund (WWF) Brazilian Biodiversity Fund, German Development Bank, Global Environment Facility and World Bank, is a 10-year project aimed at increasing protection of the Amazon. By 2008, 32 million hectares of new parks and reserves were created in the Brazilian Amazon under ARPA, among them the 3.88 million-hectare Tumucumaque Mountains National Park, one of the world's largest national parks. Growing trees binds the soil and prevents it from being washed away. Nutrients from falling leaves improve the soil and increase crop output.

<aside>
🔍 **Hint**

Avoid bullet points in answers. Listing methods will give you very few marks. Each method should be explained.
</aside>

Figure 6.21: *Afforestation*

3. **Education** can be used to explain to people the local and global consequences of cutting down trees in the rainforest. If people are aware of the negative impacts, then they are less likely to continue misusing and exploiting the rainforest. 'Sustainable development is development that meets the needs of the present, without compromising the ability of future generations to meet their own needs' (from UN Report, *Our Common Future*).

Sustainable Development

Figure 6.22: *Education*

4. **National parks and reserves** are areas created by the government and protected by laws. These laws prevent or minimise the harm caused to the landscape by commercial developments like logging and gold mining. For example, the Jau National Park is the largest rainforest reserve in South America.

Figure 6.23: *Location of the Jau National Park*

5. **Eco-tourism** is popular within national parks and reserves as tourists are attracted to the scenic views, diverse plant life and fascinating wildlife species. National parks and reserves allow tourists to enjoy these areas whilst protecting them (to some degree) against widespread destruction. In addition, the money made from tourists is invested into conservation programmes.

> **🔍 Hint**
>
> You may have studied tourism in detail for National 5 Geography.

Figure 6.24: *An eco lodge*

Figure 6.25: *Rainforest products*

6. Sustainable forestry management can be used to maintain rainforests as functional ecological environments whilst providing multiple economic benefits. A recent study by the International Tropical Timber Organisation (ITTO) found that more than 90% of tropical forests are managed poorly or not at all. Sustainable forestry management may include greater involvement of local communities, the development of plantation forests on degraded lands and non-forest, and the diversification of forest products to include non-wood products such as fruits, nuts, fragrances or seeds.

Success of methods of soil conservation/reduction of land degradation

Agroforesty has allowed small farmers to become independent and their local knowledge to be valued. The land has become more productive, more profitable and more sustainable than forestry or agricultural monocultures. It improves soil fertility and growing different crops makes it possible to harvest these crops over the year, giving an income and food supply the whole year round.

Afforestation has resulted in a significant area of forests being restored. However, the deforestation rate is still faster than the rate at which forests are restored, as incentives given to locals are unattractive. Only teak trees are planted to restore an entire area of forest, resulting in a loss of the original biodiversity of tropical rainforests. A plantation of teak trees alone cannot support the variety of flora and fauna species of the original rainforests.

In many tropical countries, forests are government owned. Private logging firms are not allowed to own the land but are given the rights to remove the timber. Without owning the land, these firms are reluctant to make investments in long-term forest management. This results in poor management of the land, thus leading to degradation. Although many tropical countries have laws to protect the land, they do not have the resources to enforce them. This means that loggers may ignore the environmental impact on the land and people. Illegal logging is a problem for countries like Brazil.

Make the link

If you take Business Management, you will have learned about sustainable production and fair trade.

Figure 6.26: *A plantation of teak trees*

The creation of national parks has been partly successful as a proportion of land is now protected, reducing degradation. However, there are not enough rangers to ensure that the rainforest is not being abused.

Figure 6.27: *A lodge in the Yasuni National Park*

Eco-tourism is more successful as degradation has been reduced in some areas. Eco-tourism can fund efforts both through park entrance fees and employing locals as guides and in the handicraft and service sectors (hotels, restaurants, drivers, boat drivers, porters, cooks). Many lodges in and around protected areas charge a daily fee to visitors which goes toward supporting the forest.

Sustainable forestry needs to involve the local people on a greater scale for it to have an impact.

Rural land use: Conflicts and their management
Cairngorms National Park case study

The Cairngorms is an area located in the north-east of Scotland. It encompasses five of the six highest peaks in Scotland, breathtaking glaciated scenery and an incredible diversity of wildlife and plants. The Cairngorms became a national park in 2003 and is the largest national park in Britain. Approximately 18 000 people live within the park boundary (shown in Figure 6.28). Although this may sound like a lot of people, the population density is actually very low due to the size and relief of the national park.

More than 1.4 million people visit the park each year. Tourism is very important within the park as it contributes significantly to the economy. Due to the accessibility and diverse landforms and landscapes within the Glenmore area, people participate in a wider range of outdoor activities than in any other similar landscape in Scotland – for example, downhill skiing in Coire Cas, cross-country skiing in the northern corries, rock and ice climbing and hillwalking. Loch Morlich has sailing, canoeing and other activities, while the general public walk and ramble in the forests and other areas. Unfortunately, these tourist activities can conflict with other land users and the local people living within the park area.

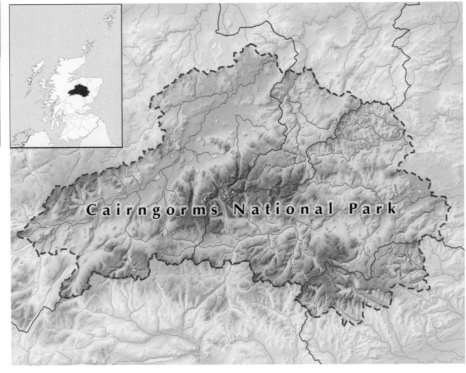

Figure 6.28: *Location of the Cairngorms National Park*

Conflicts

Traffic congestion is common in the Cairngorms area, especially in places like Aviemore, a well-known and popular holiday destination within the national park. Congestion is particularly high during the winter period when many people visit the national park to participate in winter sports such as skiing and snowboarding. For example, in recent years police were forced to close the road leading to the Glenshee ski resort as thousands of winter sports enthusiasts and sightseers flooded in to take advantage of the record snow levels. The car park at the ski resort was full and the road was in danger of being blocked by a record number of visitors.

Figure 6.29: *Congestion at Glenshee ski resort*

Congestion is also a problem on narrow rural roads and in car parks, such as the Tesco car park in Aviemore. MSP Fergus Ewing said, 'It is clear that the existing Tesco supermarket is simply inadequate for the demands placed on it and traffic congestion by the site is becoming an increasing problem.' Large volumes of visitor traffic can lead to an increase in air and noise pollution and can spoil the attraction of smaller local villages.

Many local residents believe the area's traditional roots have been lost as hotels, shops and cafes have opened and unsightly visitor and leisure complexes and accommodation have been developed. The Macdonald Aviemore Highland Resort has proposed expanding its resort to include further retail, leisure, commercial, business/office premises, holiday lodges and housing, which some people believe will have a significant visual impact on the surrounding area.

- **4 Hotels and 18 Woodland Lodges**
- **Leisure centre with pool**
- **World-class conference centre**
- **3D cinema**
- **Luxury brand shopping**
- **18-hole championship golf course**
- **Various cafes, bars and restaurants**
- **Kids club and evening entertainment**
- **Popular destination for Highland weddings**

Figure 6.30: *Amenities already offered by the Macdonald Aviemore Highland Resort*

The Cairngorms area is becoming increasingly overcrowded with tourists using these facilities. Thus, locals may decide to leave because of this. Also, the increase in holiday home ownership can push up prices, forcing many local people to move away and leaving rural areas empty during off-peak times of the year.

Some believe the natural appeal of the Cairngorms National Park is being compromised by rubbish bins, signs and man-made walkways. Activities such as skiing and hillwalking around beauty spots is causing the erosion of ski runs and footpaths. Visual pollution is also an issue, e.g. cable cars and ski tows. The funicular railway detracts from the natural setting (see Figure 6.32) and also allows more people to access the summit, thus increasing pollution and footpath erosion.

Figure 6.31: *Visual pollution caused by the Glenshee resort ski lift*

Figure 6.32: *Visual pollution caused by the funicular railway*

Visitors can cause problems for farmers and landowners within the national park, e.g. they may damage stone walls, leave gates open allowing animals to escape or disturb sheep during lambing season. A local land owner near the River Spey, says: 'Well, for a long time I was not really very happy about people crossing my land. I keep many cows and sheep and often people let their dogs off the leash and they chase my animals. They also leave my gates open and my stock can escape.'

Areas of local interest such as Glenmore Forest Park may be spoiled and rare species of plants and animals damaged. General anti-social activities may arise, e.g. litter and vandalism.

Commercial forestry also takes place within the national park. This can create conflict with both tourists and locals. Logging can cause noise pollution which disrupts the peace and quiet of the area and can scare away wildlife. The heavy trucks used to transport the logs can cause traffic congestion and increase journey times as they are slow moving. Logging scars the landscape and makes it look unsightly. Growing foreign trees in rows or lines looks out of place and destroys the natural beauty of the area. Commercial forestry can impact on tourists when walking routes and paths are closed or rerouted, affecting their enjoyment of the scenery and wildlife.

Management solutions

Traffic congestion can be reduced by introducing various restrictions, such as one-way streets, bypasses or complete closures in more popular areas and at peak times of the day/year. The Aviemore Centre ring road was constructed to minimise congestion caused by tourists on small country roads on the approach to the Cairngorms ski resort car park. Furthermore, Aviemore has restricted parking access in some residential areas by only issuing parking permits to local people.

Improved car and coach parking could also help minimise congestion in this area as it would prevent tourists from parking along the sides of narrow roads, narrowing them further and making it difficult for other cars to pass by quickly and safely.

Figure 6.33: *Vehicle parking facilities at Cairngorm Mountain base station*

The Cairngorms National Park has a strong voice in issues such as planning permission for the new developments at the Aviemore Highland Resort and setting up special areas. The national park can attach stipulations to planning applications, such as ensuring that new buildings, car parks etc. are screened behind trees and that the developments may only use local stone for buildings. This ensures that new developments, which will improve the economy, can go ahead but that they do not spoil the natural look of the landscape. This is in keeping with one of the aims of the national park – to conserve and enhance the natural beauty, wildlife and cultural heritage of the area.

Good visitor education, for example signposting, leaflets or constructing information centres, can help minimise footpath erosion in sensitive areas. Some areas that become badly damaged can be fenced off from the public altogether. Special walkways, e.g. using duckboards or 'terrain' (a supportive three-layer 'carpet'), can also be constructed to prevent footpath erosion.

National park rangers are employed within the Cairngorms National Park to promote tourists' understanding and enjoyment (including enjoyment in the form of recreation) of the special qualities of the area. They can also encourage people to visit less fragile areas and ensure tourists are aware of the importance of removing their litter so as not to spoil the scenery or harm wildlife within the park. More bins have also been set up in car parks and in the areas most frequently visited by tourists.

The law can also help prevent damage to the national park by protecting the landscape with legislation and setting up statutory bodies, such as Scottish Natural Heritage (SNH). SNH was established under the Natural Heritage (Scotland) Act 1991 and designates Sites of Special Scientific Interest (SSSIs) to protect the best of our natural heritage. There are now several SSSIs within the Cairngorms National

Park including Abernethy Forest. This forest is important for charting the vegetation history of the Cairngorms area since the end of the ice age and so must be protected.

Figure 6.34: *Abernethy Forest*

Hint

Remember! You can be asked about the effectiveness of management strategies in this section of the rural topic too.

Many of the paths are now suffering from some level of erosion. Without regular maintenance, footpath erosion can become severe and access becomes restricted. Footpaths are now being made more durable through sub-soiling. This creates a hard-wearing surface that requires less maintenance. It is designed to blend in with local stone. It has proved popular with locals and tourists as footpath erosion and scarring of the landscape has been reduced. However, it can be costly and time consuming to keep up with necessary repairs. The maintenance of mountain footpaths is becoming more important as visitor numbers increase and outdoor activities become more popular.

However, in all these issues, careful management can allow tourists to coexist with other land users.

Dorset coast case study

Dorset is an area located along the south coast of England. It contains some of the most breathtaking examples of coastal features in the UK. The Dorset coastline, or the 'Jurassic coast', is internationally recognised and was designated a World Heritage Site in 2001 by the **United Nations Educational, Scientific and Cultural Organisation** (UNESCO). Approximately 342 000 people live along the Jurassic coast (see Figure 6.35). The site is approximately 95 miles or 155 km long, and just under 1 km wide at its widest point.

Make the link

You may have learned about land use conflicts at the coast as part of the Coastal Landscapes topic in National 5 Geography.

Did you know?

The Dorset and East Devon Coast World Heritage Site is England's first and only natural World Heritage Site.

Hint

You may have studied tourism in detail for National 5 Geography.

The land within the boundaries of the World Heritage Site is looked after by its many different owners. The site is owned by over 80 separate landowners, and is used by a wide variety of people including the Ministry of Defence, local community, farmers, tourists and conservationists. All these individuals want to use the area for different activities, resulting in conflicts. The main conflicts in this area arise due to the high volume of tourists visiting Dorset in the summer months – there can be over 400 000 people visiting this area each year, with 36% visiting during the summer months and over 90% of these arriving by car. They are attracted by the awe-inspiring scenery, the climate and the variety of activities on offer in the area.

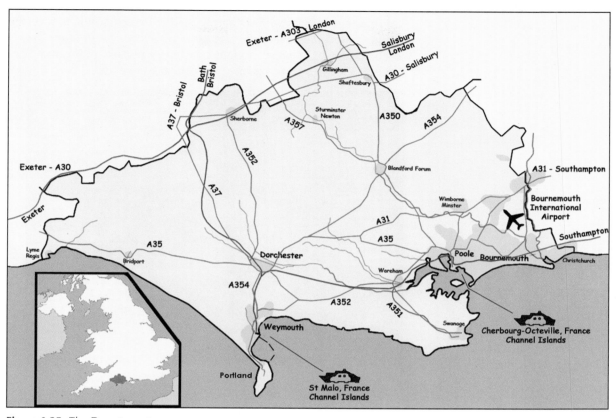

Figure 6.35: *The Dorset coast*

Conflicts

The high number of tourists arriving by car means that traffic congestion is a huge problem in the Dorset area. The car parks at Studland and Lulworth Cove have limited access so there is a concentration of cars in this area. Tourists don't always consider where they are parking and can park on grass verges or block exits, restricting access for local people and businesses. The local people also complain about the noise and litter created by the visitors, which spoil their village and its surrounding environment.

Figure 6.36: *Parking is a problem in this area*

? Did you know?

Poole Harbour is the largest natural harbour in Britain. It was formed at the end of the last ice age when what was a valley was filled by rising seas.

⁙ Make the link

You will learn more about traffic management as part of the Urban chapter.

🔍 Hint

Make sure you know a number of land use conflicts, not just those related to tourism.

The Ministry of Defence uses land behind Lulworth and Studland for training purposes as well as barracks. This causes conflicts with the tourists, especially walkers, as the MOD closes roads and coastal footpaths during exercises and the coastline is Dorset's main attraction. Access to some beaches is only possible at weekends.

Figure 6.37: *Lulworth Camp is located near the Dorset coast*

There are many important conservation areas in Dorset. These include an RSPB reserve, an Area of Outstanding Beauty, Sites of Special Scientific Interest and a Heritage Coastline. Conservationists feel that the large numbers of tourists using the area, especially during the summer months, are causing environmental damage.

Figure 6.38: *The area around the village of Arne near Swanage is an RSPB nature reserve and is popular with walkers and bird watchers*

The coastal footpaths are being deeply eroded, creating eyesores on the coastline. This is particularly obvious on the path from Lulworth to Durdle Door – one of the most heavily walked paths in Britain. Prevention methods can result in access being limited in some places, causing conflict with tourists.

> This path has been fenced off to prevent erosion on the cliff edge. Thank you for your patience and care in this matter. If you have any queries regarding the footpath please contact Dorset Countryside.

Figure 6.39: *Signpost on Dorset coast*

Farmers also come into conflict with tourists who walk across farmland, leave gates open and restrict access to fields by irresponsible parking.

Around the coast in areas like Poole Harbour there are many different users of the coast. There can be up to 4500 boats in the area in high season using the harbours, marinas and sea areas. Competing with the boats for use of the coastal area are fishermen and anglers. They can be disturbed by boat wakes as well as jet skis and water skiers. Many tourists are attracted by the lovely beaches for swimming and sunbathing but can be disturbed by the noise of motor boats.

Management solutions

Figure 6.40: *Car parking, Studland*

Around the Studland area, the four main car parks have been expanded and can now accommodate another 820 cars so as to stop cars parking on grass verges and other inappropriate parking. However, some visitors still do not use the car parks and continue to damage the verges and block farmers' field gates.

Some paths have been closed to prevent further erosion and sand dunes have been fenced off. The fences successfully collect sand where trampling or wind erosion has taken hold. Boardwalks are laid along main footpaths to reduce footpath erosion. To reduce litter, bins are put along the paths and at the back of beaches. The walkways are designed on several levels, using a variety of paving, cobbles and stone work which complements the appearance of the buildings and other structures within the adjacent areas. However, some tourists still leave litter behind. Fires are started by irresponsible tourists, so fire beaters are available and in some areas fire breaks have been formed.

Figure 6.41: *Poole Harbour*

Zoning of areas ensures that different activities like fishing and jet skiing are kept apart, e.g. at Poole Harbour. Speed limits have been put in place to reduce the impact of boat wakes and cause less disruption to beach users and coastal erosion.

The public are being educated on how to look after the environment through the use of leaflets and information boards. This is proving successful as more people are aware of the need to look after the countryside. Rangers are also available to give advice to the public on protection of the environment. Organisations like the National Trust and English Nature have taken on some of the responsibility for looking after the coastline. In some areas a fee is charged by the National Trust for the facilities provided and this is used to further protect the area and for implementing improvements to the beach and its facilities.

Summary

In this chapter you have learned:

- The impact on the people and the environment of land degradation in a semi-arid area (Sahel)
- Strategies to reduce degradation
- The impact on the people and the environment of land degradation in a rainforest environment (the Amazon)
- Strategies to reduce degradation
- Success of the strategies
- Land use conflicts and their management related to a glaciated area – case study of the Cairngorms National Park
- Land use conflicts and their management related to a coastal area – case study of the Dorset coast
- The success of management strategies for glaciated and coastal areas.

You should have developed skills and be able to:

- Identify semi-arid and rainforest areas from a map
- Analyse solutions to degradation from a variety of sources.

End of chapter questions and activities

Quick questions

1. For a named area you have studied, discuss the consequences of rural land degradation on the people and the environment.

2. Evaluate the effectiveness of these measures.

3. Look at the table below. For either a semi-arid area or the Amazon Basin, explain measures used to prevent rural land degradation.

Soil conservation strategies	
Semi-arid area	**The Amazon Basin**
Animal fences	Agroforestry
Dams built in gullies	Crop rotation
Stabilisation of dunes	Return to traditional farming
'Magic stones' (diguettes)	Purchase by conservation groups

4. With reference to the Cairngorms National Park, explain the land use conflicts that may occur.

5. Explain, in detail, the methods used to manage these conflicts.

6. Analyse how successful these strategies have been.

7. With reference to the Dorset area, explain the land use conflicts that may occur.

8. Explain, in detail, the methods used to manage these conflicts.

9. Explain, in detail, how successful these strategies have been.

Exam-style questions

1. Referring to named locations in either a semi-arid area or a rainforest area in a developing country, **discuss** the consequences of rural land degradation on the people and their environment.
   8 marks

2. Referring to named locations in either a semi-arid area or a rainforest area in a developing country:

   a. explain soil conservation strategies aimed at reducing land degradation.

   b. comment on the effectiveness of these strategies.
   10 marks

3. For any named upland or coastal area you have studied:

    a. discuss the environmental conflicts that may be caused by large numbers of people visiting the area for tourism and recreation.

    8 marks

    b. discuss the measures taken to resolve these environmental conflicts and evaluate their effectiveness.

    10 marks

Activity 1: Group presentation

The government of Brazil wants to develop part of its rainforest. Possible developments include mining, cattle ranching, HEP and commercial logging. However, the government is concerned that the people and the land in the area selected should be protected from degradation. Any company bidding to develop the land must include a proposal on how they will achieve this aim.

There are representatives from the mining industry, cattle ranchers, an electricity company and loggers bidding to develop the forest. Planning is dependent on strategies being put in place to protect the people and environment of the area. The successful bid will provide the best outcome for the area.

Divide into groups of four. Each person should choose to represent one of the developers listed.

For your chosen development, use the internet to prepare a report on the strategies you would put in place to protect the people and environment of the area to stop land degradation.

Share your information with your group. Using the reports from your group, create a poster/newscast/PowerPoint to record the group's findings.

Activity 2: Revision cards

In pairs, collect small pieces of card and make up revision cards. Use the glossary for this chapter and write each word on the front of the card and its definition on the back. Test each other to see if you can explain each word.

Learning Checklist

You have now completed the Rural chapter. Complete a self-evaluation sheet to assess what you have understood. Use traffic lights to help you make up a revision plan to help you improve in the areas you have identified as amber or red.

- Account for the main causes of land degradation in the Sahel.

- Discuss the main human consequences of land degradation in the Sahel.

- Discuss the main environmental consequences of land degradation in the Sahel.

- Discuss the methods of soil conservation used in the Sahel/semi-arid areas.

- Evaluate the success of these methods.

- Explain the main causes of land degradation in the Amazon rainforest.

- Discuss the main human consequences of land degradation in the Amazon rainforest.

- Discuss the main environmental consequences of land degradation in the Amazon rainforest.

- Discuss the methods of soil conservation in the Amazon rainforest.

- Evaluate the success of these methods.

- Explain the main land use conflicts in the Cairngorms National Park.

- Discuss solutions/management/effectiveness of these conflicts.

- Explain the main land use conflicts on the Dorset coast.

- Discuss solutions/management/effectiveness of these conflicts.

Glossary

Afforestation: deliberate planting of trees on otherwise bare land.

Aid: the giving of resources by one country or organisation to another country.

Biodiversity: the variety of plants and animals in the world.

Cash crop: crops grown for the purpose of generating income.

Climate: average weather conditions over at least 35 years.

Climate change: the long-term alteration of weather patterns over time.

Contour ploughing: the practice of ploughing along the contours of a slope in order to minimise the down-slope run-off of water and thereby prevent soil erosion.

Crop rotation: a method of farming that avoids growing the same crop in a field continuously. A regular change of crops maintains soil fertility and reduces the risk of pests and diseases.

Degradation: the deterioration of an area over time.

Desertification: the spread of desert conditions in arid regions due to human activities, drought or climate change.

Exploitation: when the natural environment is destroyed for its natural resources, e.g. deforestation.

Extinction: when a type of plant or animal is wiped out forever.

Fallow: a field left with just grass for a period in order for it to naturally regain its nutrients after several years of crops. This is usually part of a crop rotation cycle.

Famine: a shortage of food causing malnutrition and hunger.

Fertile soil: a soil that is rich in nutrients.

Footpath erosion: when ground is worn away.

Greenhouse effect: the retention of heat in the atmosphere caused by the build-up of greenhouse gases.

Indigenous people: native tribes.

Irrigation: artificially watering crops from stored water.

Jurassic Coast: the name given to the coastline around Dorset.

Land use conflict: when there are conflicting views on how land should be used.

Migration: movement of people from one area to another.

MOD: Ministry of Defence.

Monoculture: a farming system in which a single crop is grown continuously in the same field. This can exhaust the soil nutrients, lead to a breakdown in soil structure and the loss of soil through wind or rainwater erosion.

National park: a protected area created to conserve and enhance the natural beauty, wildlife and cultural heritage.

Nutrient cycle: how the minerals that provide energy and nutrition to living organisms circulate around an ecosystem.

Nutrients: chemical elements that are essential for plant nutrition.

Over-cultivation: the excessive use of farmland to the point where productivity falls due to soil exhaustion or land degradation.

Over-grazing: the destruction of the protective vegetation cover by having too many animals grazing upon it.

Planning permission: permission that is needed to build or change buildings.

Pollution: air, noise or visual disturbance of a local environment.

Ranching: rearing of beef cattle on a large scale.

Selective logging: when trees are only cut down once they reach a certain height.

Soil erosion: where earth is washed or blown away.

Soil exhaustion: where the soil has lost its fertility.

Strip cropping: where crops are grown in narrow strips; the ploughed soil where seeds are to be planted is protected from erosion by adjacent strips where growing crops are at much later stages of growth.

Sustainable development: development that meets the needs of a population without polluting the environment or depleting natural resources.

Sustainable farming: farming that avoids soil erosion and pollution; it does not destroy the land for future generations.

Sustainable logging: controlled removal of trees, allowing regrowth.

Tourism: when people travel to places for pleasure.

Traffic congestion: occurs when too much traffic tries to access a particular area at the same time.

7 Urban

Glasgow case study

Background

 Hint

The name 'Glasgow' comes from the Gaelic meaning 'green valley' or 'dear green place'.

The original site of Glasgow was on the banks of the Molindar burn and at a natural ford on the river. The terraces of the river provided early settlers with flat land to build on and to grow crops. The River Clyde also provided food and water and the ford allowed it to develop as a trade centre for people travelling from other areas of Scotland. Situated on the west coast, Glasgow was in an ideal location to trade with the American colonies and the raw materials imported encouraged the growth of the textile and tobacco industries. This, alongside the industrial revolution, resulted in a large influx of people looking for work. In the 1950s, Glasgow's population rose to over one million. Today Glasgow is Scotland's largest city with a population of 611 748 in July 2019.

Make the link

If you study History, you may have learned about industry and migration in Glasgow over the years.

Figure 7.1: *Glasgow city centre*

Tenement housing

During the 1800s, the development of the shipbuilding industry in Glasgow, and its increasing need for workers, caused an enormous rise in the city's population.

This saw the rapid development of tenement blocks to house the workers. Between 1900 and 1951, the population of the city continued to rise – reaching a peak of 1.1 million in 1951. The tenement flats became slums and Glasgow was one of the poorest cities in Europe.

Figure 7.2: *Glasgow in 1939*

The flats were very small and the rooms were usually dark and damp. They were often overcrowded with families of up to 10 cramped in one or two rooms. The overcrowding and damp conditions caused disease, such as tuberculosis and bronchitis, to spread quickly. Due to their close proximity to the factories where most residents worked, people were often exposed to smoke and chemicals, causing further diseases like asthma. The houses lacked basic amenities such as electricity, running water, central heating and even indoor toilets. The entire tenement often shared only one or two toilets on the landing or an outhouse was found at the back of the building. In the tenement closes it was common to find rats living among the filth. Due to the poor conditions in the tenements, life expectancy in areas like the Gorbals was often no more than 50 years.

Figure 7.4: *Problems with the tenements*

Figure 7.3: *The tenements*

:: Make the link

If you take Modern Studies, you may have looked at deprivation in Glasgow in detail.

○ Hint

The information on tenement housing is good background knowledge. Because the topic is on recent changes, you should concentrate on later problems.

? Did you know?

In 1840s Britain, many urban households dealt with waste disposal by keeping a pig. The pigs would eat toilet waste – and families would sometimes then eat them.

After the Second World War, it was evident that Glasgow faced major housing problems. The tenements were in a terrible condition and were simply not safe for people to live in, so the decision was taken to move some residents to council estates on the edge of the city or new high-rise flats.

Problems in the council estates

When people were moved from the tenements to council estates on the edge of the city, the close-knit community spirit of the old inner-city areas was destroyed. When building these council estates, the planners forgot entirely to include space for local amenities like shops, pubs, cinemas and community buildings. Neither did they provide any places for people to work. Most of these areas only had a post office, one telephone box and a half-hourly bus service to the city centre. This meant that those without cars had very few ways to travel out of the area to access the amenities they needed. People were left feeling isolated and unhappy in these new areas and so the estates became run down and vandalised. There was high unemployment among the residents and crime rates rose dramatically. Families chose to leave the area and many houses were abandoned and fell into disrepair.

Make the link

You may have looked at how crime is linked to factors such as unemployment and geographical location in Modern Studies.

Figure 7.5: *Original Corporation housing, now derelict, in Lochend Road, Easterhouse*

Hint

Make sure that you learn **named examples** of case studies as you will gain marks for demonstrating your knowledge, especially on the most recent changes taking place in your chosen cities.

Problems of the high-rise flats

The high-rise tower blocks were no better. A 'build-'em-high, build-'em-quick' approach saw hundreds of high-rise flats constructed across the city. In the rush to build these high-rise flats, poor materials were used which later led to structural issues. Due to their flat roofs and the often rainy Scottish climate, dampness was common. Similar social problems to those of the council estates occurred, e.g. people felt isolated, especially those living on the upper floors when the lifts were out of service or vandalised. Again, families chose to leave the high rises and many flats were left empty, which attracted criminals, drug dealers and squatters.

Figure 7.6: *Tower blocks*

Housing management strategies employed

Comprehensive development areas

Given the worsening living conditions in the tenement houses of Glasgow, the city council made the decision to bulldoze large areas of tenements and build better housing for the inner-city residents. Glasgow called these areas CDAs – comprehensive development areas. In total, 29 inner-city slum areas were designated CDAs, including parts of Govan, Partick, Springburn and the Gorbals.

Knocking down the tenements meant that the residents needed to be housed elsewhere. Glasgow moved these people into outer-city council estates and high-rise flats as mentioned above. However, some were also moved into specially designed new towns or allowed to return to tenements that had been renovated instead of demolished.

The council estates

Glasgow established four main council estates for over 200 000 people from the inner-city during the 1950s and 60s – Castlemilk, Drumchapel, Easterhouse and Pollok. These areas were originally planned as low-density, semi-detached houses, with gardens and set in attractive countryside. However, cheaper three- and four-storey buildings of flats were eventually built instead as it was realised that the original plans would cost the council too much money and take up too much land.

Figure 7.7a: *1950s corporation housing in Pollok*

The new towns

There were three new towns built around Glasgow during the 1950s and 1960s to house the overspill from the inner cities. East Kilbride is the largest of these towns with a population of 75 120 in 2019.

Unlike the council estates, the new towns were thoroughly planned settlements, designed to incorporate workplaces and services such as shops, libraries, schools and leisure centres for residents. Foreign companies were encouraged to locate in industrial estates on the edge of the new towns. They were offered incentives such as loans or grants and new buildings so that they would provide jobs for residents in the area. Houses were a mix of flats, cottages, semi-detached and detached housing to attract a range of people. East Kilbride's well-known road network of roundabouts and few traffic lights was typical of new towns to help keep traffic flowing quickly and easily.

Figure 7.7b: *Roundabouts are common in Scotland's new towns*

Tenement renovation

Not all the tenements had been demolished in the CDAs so the decision was made in the late 1970s to renovate those that remained. Several flats were often combined to make one larger flat. Electricity was installed, along with central heating, double glazing and internal bathrooms. The outer walls were cleaned to remove the pollution that had built up from the old industries and the surrounding areas were landscaped to look more pleasant. Streets were often made into one-way systems to reduce traffic and make the areas safer for families with children.

Redevelopment areas
Crown Street Regeneration project

The Gorbals, situated just south of Glasgow city centre, was part of the post-war redevelopment scheme. Originally it was designed as a suburb for the middle classes but they chose to live in the west of the city rather than the south. Industry developed in the area and low-quality tenement housing was built to house the workers. Workers poured in and it became one of the most overcrowded slum areas in Scotland. The houses were cramped, damp, suffered from pollution and had no inside toilets. In the 1930s, the area developed a gang culture and became known as one of the roughest areas in Glasgow, if not the UK.

Post-war regeneration aimed to tackle the area's problems. Much of the housing was demolished and replaced with high-rise flats. Hutchesontown was built in 1968 but the houses were poorly built and damp so in 1982, only 14 years after completion, the last residents moved out of their homes. Due to the failure of this housing to regenerate the area it was decided to demolish the scheme in the late 1980s. The Crown Street Regeneration Project was set up in 1990 to revitalise the area. The aim at Crown Street was to regenerate the area and create a place where people would want to live, where people could enjoy the privacy of their home but still be part of the local community. The area contained local amenities like shops, landscaped areas to improve the look of the area, safe areas like playparks for children to play as well as the provision to own or rent your home.

Figure 7.8: *Regeneration in the Gorbals*

Red Road flats

Built in the mid-1960s to tackle the city's housing crisis, the flats once provided accommodation for almost 5000 people. There were eight tower blocks in total and when they were built, at 292 ft (89 m), they were the tallest residential structures in Europe. They were intended to be a fast and cost-effective solution to the problem of overcrowding but in recent years have become rundown and vandalised, and stand largely empty. In 2003, ownership of the flats was passed to Glasgow Housing Association. However, it was found that ongoing repairs were costing more than receipts in rent, so in 2005 Glasgow Housing Association decided to demolish one of the tallest blocks as part of a regeneration plan for the area. Regeneration is well underway with new housing developments and the demolition process of the flats ongoing. It was proposed to demolish one of the towers as part of the opening ceremony of the 2014 Commonwealth Games but after objections it did not go ahead.

Figure 7.9: *Red Road flats, 1969*

Figure 7.10: *Red Road flats, 2014*

Glasgow Harbour development – 21st century

Glasgow Harbour is a 130-acre, high-quality mixed-use development on the west end of Glasgow's water front. This brownfield site was originally used for the loading and unloading of ships with warehouses, storage areas, etc. The need for bigger ships meant that this area was no longer suitable and became derelict. The owners, along with Glasgow City Council, created a project to develop the site. There are 2500 apartments, offices and a retail and leisure district with shops, bars, restaurants, cafes, health and fitness facilities as well as the city's new Riverside Museum. There are acres of parks and public space and riverside walkways giving access to over 3 km of waterfront.

Glasgow Harbour is different from the Gorbals as the housing is a completely different style as well as the majority of housing being privately owned, and building density is much higher. Glasgow has a lack of premium quality housing, a need that Glasgow Harbour is addressing.

Figure 7.11: *Glasgow Harbour development*

Pacific Quay

Across the river from Glasgow Harbour, the Pacific Quay development is a mixture of business, housing and leisure. It is situated on the site of Glasgow's Princes Dock, which closed during the 1970s. It includes a media village, leisure developments, office space, several hotels and over 300 new houses. This provided much-needed additional housing stock.

Housing in the east end of Glasgow, 2014

In the east end of Glasgow, a 33-hectare site on Springfield Road was developed to create an athletes' village for the 2014 Commonwealth Games. The village provided accommodation and facilities for 6500 athletes and officials for the duration of the Glasgow Games. It was planned that after the Games, some of the housing would become social housing while some would be for sale. A total of 300 of the houses and flats are for sale by the developer, and 400 houses are available for rent. They are a mixture of two-, three- and four-bedroom family houses, all with gardens. A number of them will be fully wheelchair accessible. The houses are designed to be more energy-efficient, thus reducing bills as well as climate change.

Gentrification

Gentrification occurs when run-down areas of the inner-city or CBD are restored and renovated by wealthy groups moving back into these areas to live. Property prices increase as a result. Several parts of the centre of Glasgow have been gentrified, such as the Merchant City. Glasgow's Merchant City is found at the eastern edge of the CBD. During the 18th and early 19th centuries, this was the main trading area. This area became more and more run down until by the 1980s many buildings and warehouses lay derelict. Significant amounts of money have been spent restoring traditional buildings and converting them into flats, offices, hotels, bars and restaurants. Examples of gentrification include the conversion of the former *Evening Times* newspaper offices into luxury flats.

Make the link

You may have looked at the impact of the 2014 Commonwealth Games on Glasgow in Business Management, or PE.

Figure 7.12: *Example of housing in the athletes' village*

Impact of housing management strategies

Due to the council estates falling into disrepair and the vandalism they experienced, Glasgow City Council has been forced to spend a huge amount of money improving these areas over the last few decades. They renovated the existing buildings to improve living conditions and installed CCTV cameras to try to prevent vandalism from reoccurring. In areas such as Pollok, the buildings were in such poor condition that they could not be renovated and had to be knocked down entirely. The high-rise flats, like the Red Road flats, faced the same fate and today almost all of these have been demolished. New housing developments are now being built on the land. The ones that remain have been modernised and security measures increased. Controlled entry systems have been installed and wardens patrol the blocks to improve safety. Renovating the tenements turned out to be a much cheaper and more effective option than building new council estates or high-rise flats.

The Gorbals renovation scheme has achieved its aim of creating a place where people want to be. The area has a thriving local centre and most of the residential units are occupied. The main streets are lively and the public spaces are well maintained. The basis of the development was Glasgow's tenement tradition. It has provided a clear separation between public space and the semi-private space in the courtyards. The residents were involved in the project from the beginning and have taken ownership of the scheme. They actively maintain the area and their commitment has enabled the further regeneration of the adjacent Queen Elizabeth Square and Moffat Gardens areas to the east and the Laurieston area to the west. Artwork reflecting the social history of the area is displayed in various ways in the area. This involved local artists and gave the people a sense of pride in their new environment.

Glasgow Harbour has been partly successful as it has provided premium quality housing, which Glasgow previously lacked, whilst regenerating a derelict area. However, the housing in many cases is too expensive for the original local population and they have been forced to move away from the area.

In the east end, the athletes' village has improved the availability and standard of housing in the area.

Many housing areas in new towns, like East Kilbride, are now highly sought after. Their outskirt locations make these areas attractive to live in, offering a quieter, less hectic environment than the inner city while still offering all the required amenities. Improved transportation routes into the city, e.g. the M77 and the M74 extension, allow easy access to the city for those who work there.

Gentrification

Although the buildings in the area are restored to their original states, new housing is created and crime is reduced, gentrification can cause problems. The huge increases in property prices often force out the traditional working-class people who previously lived in these areas, as they simply cannot afford to buy or rent property in these areas. The people who remain may also be unhappy at these changes and resent 'outsiders' moving in.

> ### ☄ Make the link
> You may have learned about how CCTV is used to tackle crime in Modern Studies.

> ### ☄ Make the link
> You may have learned about community art projects in Art, or in Modern Studies.

Need for transport management

Glasgow, like many large cities, suffers from problems caused by increased traffic. Traffic congestion in Glasgow is a result of the increased volume of cars on the road today. In 2000, the Department for Transport (DfT) recorded 978 317 cars on major roads in Glasgow. By 2013, this number had risen to 1 113 695, an increase of 135 000. It is clear that more people are choosing to travel by car as it is more convenient than public transport.

Glasgow City traffic profile for 2000 to 2013

Year	2000	2001	2002	2003	2004	2005	2006
Cars	978 317	992 442	1 016 013	1 003 637	1 044 912	1 054 158	1 061 247
All motor vehicles	1 203 103	1 222 686	1 248 898	1 243 415	1 294 023	1 315 126	1 329 566

Year	2007	2008	2009	2010	2011	2012	2013
Cars	1 067 007	1 069 907	1 092 601	1 075 541	1 077 737	1 090 962	1 113 695
All motor vehicles	1 339 233	1 344 411	1 355 448	1 341 893	1 348 197	1 369 817	1 404 549

According to the 2011 census, a total of 11 260 336 working residents in the UK commuted from one local authority to another for work. In Glasgow approximately 25% of all people who work in the city commute to work. As shown in Figure 7.13, the largest percentage of these commuters (32.6%) make their journey to work by driving in either a car or a van, and a further 20.3% of people travel by bus, minibus or coach. This huge volume of people enters the city centre daily and causes massive congestion, particularly at morning and evening rush hours.

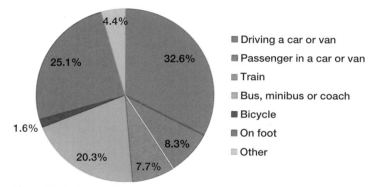

Figure 7.13: *Types of journeys*

Due to Glasgow's location on the River Clyde, traffic is funnelled across the river at very few bridging points. The Clyde Tunnel links north and south Glasgow with an average of 65 000 vehicles per day travelling through it. This funnelling creates massive congestion as people queue to pass through the tunnel or cross bridges, such as the Albert Bridge which links the Saltmarket in the city centre with Crown Street on the southside, in both directions.

Figure 7.14: *Horse-drawn carts in George Square, circa 1900*

Like many other cities in the UK, Glasgow's Central Business District (CBD) was originally built in a grid iron pattern as it was designed to cope with the horse and cart, not cars. The streets are narrow in most places, making it difficult for cars to pass each other, slowing traffic down and creating congestion in the CBD. This situation is worsened by cars parking at the side of these roads, narrowing them even further.

With so much traffic in the CBD, there are associated problems of noise and air pollution, increased journey times leading to road rage and accidents, and the damaging effects on buildings caused by the vibrations from traffic.

> ☄ **Make the link**
>
> You will learn about the effects of pollution caused by cars in the Global climate change chapter.

Transport management strategies employed

Figure 7.15: *The M8 during rush hour*

Motorways such as the M8 and the M77 were designed to keep traffic flowing quickly through the city centre. Each motorway, with up to three lanes in each direction, allowed a far greater volume of traffic to pass through the city without clogging up the small, narrow city centre roads. Expressways, such as the Clydeside Expressway near the SECC, have also improved access into the city centre. The M74 extension opened in June 2011 with the hope that it would relieve some of the traffic congestion that had begun to build up on the M8 – one of the busiest motorways in Britain.

Glasgow has recently undertaken massive modernisation in many city centre train stations such as Queen Street, Argyle Street and Glasgow Central in a bid to encourage more people to use public transport and leave their cars at home. This has involved extending some platforms in order to cope with longer trains that carry more passengers. Lines have also been extended, e.g. the line connecting Larkhall to Glasgow Central via Hamilton Central was opened in December 2005 by First Minister Jack McConnell.

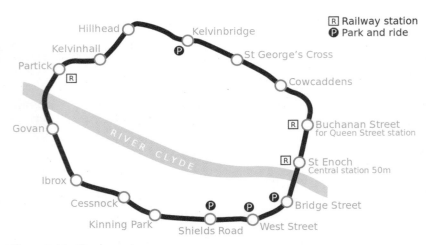

Figure 7.16: *Glasgow subway system*

Glasgow's subway system (underground railway) has also been modernised. The subway was originally run using a cable pulley system but was later electrified to increase efficiency. Its 15 stations have also been enlarged to allow more passengers to make use of it and provide a fast, efficient service to selected areas of the CBD and Glasgow's West End. Plans for a multi-million extension of Glasgow's subway system into the East End have been proposed numerous times, most recently in the run up to the 2014 Commonwealth Games.

Park and ride schemes have been introduced across the city, e.g. at Bridge Street, Kelvinbridge and Shields Road subway stations, where commuters can park their cars for free next to the station and complete their journey into the city centre by subway, thus reducing car numbers in the CBD.

Other management methods that have been put in place in the city centre include:

- Converting city centre streets to one-way systems to help traffic flow, e.g. Hope Street.
- Bus lanes have been created to allow public transport to move more quickly and cars are fined if they are caught travelling in these lanes at restricted times, e.g. North Hanover Street and Queen Street station facing north. Glasgow added a further five enforcement cameras to the existing 11 in 2013 to prevent abuse of bus lanes.
- Traffic wardens, parking meters and yellow lines prevent cars from parking at the sides of the narrow city centre roads.
- Building more multi-storey car parks, especially near shopping centres like Buchanan Galleries and the St Enoch Centre, also minimises kerbside parking.
- Pedestrianisation has helped to improve safety for shoppers by keeping cars out of the major shopping streets, e.g. Sauchiehall Street, Buchanan Street and Argyle Street.

Figure 7.17: *Pedestrianised shopping area in Glasgow*

Success of transport management strategies

The M74 extension was controversial when it was first proposed as people in the surrounding areas were required to leave their homes and close their businesses to make way for the construction of the road. Rutherglen MSP James Kelly said: 'While the overall completion of the M74 is to be welcomed, we must not forget the groups and businesses that have been displaced by the development.' The communities were also concerned about the impact of noise and air pollution from the increased number of cars passing through their areas. However, it saves around 15 minutes on journey times for through traffic as well as removing approximately 20 000 vehicles from the M8 between Baillieston and Charing Cross. In addition, up to 15% of traffic from local roads is reduced, such as at Rutherglen Main Street.

Extending the railway line to link Glasgow International Airport with the city centre stations was proposed in the early 2000s. The **Glasgow Airport Rail Link**, or GARL, was intended to have a service of four trains per hour via Paisley Gilmour Street railway station. However, on 17 September 2009 the rail link was cancelled as part of public spending cuts due to its large expense. A similar situation occurred with the proposed extension to the Glasgow subway system into the East End.

In recent years there has been a public outcry as the number of bus lanes in Glasgow city centre further restricts the movement of cars through the CBD. The public were further incensed when it was revealed that the bus gate at Nelson Mandela Place in Glasgow city centre, which imposes a daytime ban on vehicles using the bus lane, had earned the council at least £800 000 in only two months. Drivers who breached the restriction faced a £30 fine which increased to £60 if it was not paid within 14 days. The public complained that the signage indicating the restriction was not clear as often it was blocked from view by buses or other large vehicles and therefore they were fined unfairly.

> **? Did you know?**
> The M74 extension cost £692 million or £200 an inch.

Figure 7.18: *Nelson Mandela Place, Glasgow*

Rio de Janeiro case study

Background

Rio de Janeiro developed because of its natural assets of fertile land and location; sugar cane was the main crop, replaced over time by coffee and then cattle. Its location beside the sea meant a port could be developed for trading purposes. The growth of railways in the 1880s allowed the city to develop further; coffee growers could transport and export coffee with ease and it also encouraged the arrival of people from the countryside who settled along the railway tracks. By 1920, the city was becoming an important industrial centre with a population that exceeded one million, and by 1940, Rio had grown to nearly two million people. Skyscrapers and large apartment buildings replaced homes and small buildings, and poorer residents were pushed to the outer edges of the city. Rio grew larger as Brazilians without jobs or education continued to move into the city and a lack of affordable accommodation led to the building of huge favelas, or 'shanty towns', creating massive social problems for the city such as crime, overcrowding and pollution. The government is now trying to address these problems and migration from the countryside is slowing down. Today, Rio is the second largest city in Brazil with a population of six million people.

> ### Make the link
>
> You may have learned about Rio de Janeiro's favelas as part of the National 5 Geography course.

Figure 7.19: *Rio de Janeiro*

Need for housing management

Millions of people in Brazil have migrated from poor rural areas to the larger cities like Rio de Janeiro hoping for a better quality of life and better job opportunities. Unfortunately, the migrants are often unskilled and poorly educated and as a result are not able to find well-paid work. Therefore, they cannot afford the available housing. In desperation they resort to living in illegal settlements, known as favelas, on Rio's steep hillsides or on unused land near rubbish dumps or swamps. They build their homes from basic materials like corrugated iron, cardboard and broken bricks – whatever they can find. Rio de Janeiro had an estimated population of 6.4 million in 2019 and of this approximately 1.4 million were living in favelas. It is estimated that by 2025 the population of the city will grow to at least 13.1 million as more and more people migrate into the city.

> **⚡ Make the link**
>
> You will have learned more about rural–urban migration in the Population and Rural chapters.

Figure 7.20: *Favela covering the hillside of Rio de Janeiro*

The favelas are often unpleasant places to live as they experience many of the problems discussed below:

Figure 7.21: *Problems in the favelas*

> **⚡ Make the link**
>
> In the Development and health chapter you will learn more about the situation in developing countries.

The migrants build their own homes from any materials they can find. This means that they are not always very safe and they **lack basic services** such as clean running water, electricity and sewerage systems. There are open sewers running through the makeshift, narrow streets which often contaminate any clean water supplies in the area. Electricity is often 'bootlegged' (stolen) from overhead pylons, which often leads to fires and a high risk of people being electrocuted.

> ### 🔍 Hint
> Remember to give reasons for the problems in shanty towns, not descriptions.

Figure 7.22: *Massive entanglements of bootleg electrical and phone wire connections in the favela, Roçinha*

The favelas of Rio are incredibly **overcrowded**, with families of up to 10 living in one or two rooms with one bed for the adults and the children sleeping on the floor. Due to the open sewers and poor sanitation, life expectancy in the favelas is low – around 55 years. Illnesses such as bronchitis are common and **diseases** like cholera and diarrhoea spread quickly. Although medical care is available, it comes at a high cost for those living in these poorer areas of the city. Most cannot afford the cost of essential medicines, resulting in a higher infant mortality rate – 20 in every 1000 people will not reach their first birthday.

Most of the jobs held by those in the favelas are simple jobs that do not require much education. Employment is difficult to come by, **unemployment** rates are around 20% and most people who are employed are construction workers, maids, bus drivers, cashiers or hotel and restaurant workers. They are often poorly paid and there is no guarantee of work on a day-to-day basis. As favela residents receive poor education, they are not able to pass the difficult entrance exam (called the Vestibular) to get into university. Only the affluent, who can afford tutors, can gain entry. It is rare for a favela resident to get into university and, therefore, there is little opportunity to find high-paying employment and work their way out of poverty in the favelas.

Criminal gangs control many of the favelas in Rio, and drug dealing, gun trafficking, kidnapping and murder are common, everyday occurrences. More than 6000 people are killed each year, a rate similar to a war zone. As the favelas are illegal settlements, the police are not often seen as they view favelas as out of their jurisdiction.

When they are there, they are usually in full protective body armour and heavily armed.

**Make the link**

If you take Modern Studies, you will have looked at crime and gangs in Scotland.

Figure 7.23: *Traffickers sell drugs in the Antares slum in Rio de Janeiro*

People die every year in **mudslides** caused by heavy rainfall on Rio's steep hillsides. In April 2010, the heaviest rain in decades was recorded and a landslide was triggered in Niteroi, a favela across the bay from Rio. Over 200 people lost their lives as a result and at least 50 homes were destroyed. Most of the victims were the residents of the favela built on the hills. Rock falls also pose a real danger to the residents as many homes have been built below overhangs that are prone to breaking off due to erosion.

**Make the link**

You learned about erosion and the effects of heavy rainfall in the Lithosphere and Hydrosphere chapters.

Figure 7.24: *Landslide*

Housing management strategies employed

The Brazilian government originally planned to completely demolish and clear the favelas, but this method was considered ineffective in the long run as it did not solve the problem of where to house the people who lived there. Instead, it was felt that improving the living conditions for those in the favelas would be preferable.

Figure 7.25: *Housing management strategies*

The Brazilian government cannot afford to build new housing areas for the millions of people living in the favelas but they can provide them with the basic materials to help them improve their own homes and neighbourhoods. These are called **self-help schemes** and are small projects that set out to use the existing skills of the people who live within the favelas. This is the bottom-up approach.

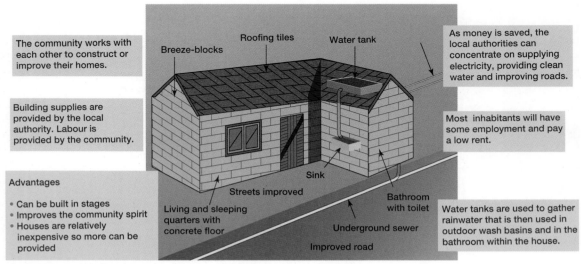

Figure 7.26: *Self-help schemes*

On the other hand, there is the top-down approach – **site and service schemes**. The government is also trying to upgrade the living conditions in the favelas by introducing basic services, like paving the roads to allow refuse lorries to collect rubbish, establishing appropriate sewerage systems and installing electricity and street lighting. Slum upgrading also often involves legalising land ownership for the residents so they are no longer considered squatters. Other facilities, such as schools and health centres, are also built to transform the favela into an established community, e.g. the Favela Bairro (Slum to Neighbourhood) Project in Rio. Rio's government has also built aqueducts to channel the water away from favela homes to help to reduce the possibility of landslides.

Projeto Cingapura was very similar to the building of high-rise flats to accommodate the industrial workers living in Glasgow's slums after the Second World War. The Brazilian government cleared large areas of favelas and built tower blocks in their place to rehouse the favela residents. The new 'superblocks' cost almost 10 times more than the site and service schemes.

Non-governmental organisations (NGOs), like Project Favela, also help to improve life in the favelas. They rely on international volunteers to help educate and mentor favela residents to give them the opportunities they need to improve their living conditions. NGOs also help people find jobs, provide medical care and build schools for the children of the favelas who cannot afford to attend the private schools in Rio de Janeiro.

The **Pacification** campaign was established in 2008 to deal with the widespread crime in the favelas of Rio. The aim was for police to take back control of the favelas and begin improving the quality of life for residents. The government gave drug dealers and gun traffickers advanced warning to leave the favelas before military police units trained in urban warfare were sent in. The police reclaimed the favela street-by-street, searching houses, cars and suspects for drugs and weapons. After the gangs are forced out, the police restore order and make it safe for town planners, social workers, etc. to begin establishing social and economic development plans, like those discussed above, to improve living conditions.

Figure 7.27: *Pacification campaign*

> ### 🔍 Hint
>
> Make sure that you learn **named examples** of case studies as you will gain marks for demonstrating your knowledge, especially on the most recent changes taking place in your chosen cities.

Hint

In the exam, make sure you read the question carefully. It may ask for the causes and effects, or perhaps about strategies and effectiveness, or any combination of these. Make sure you answer the question asked!

Success of housing management strategies

Thanks to self-help schemes, almost all of the housing in favelas like Roçinha has been improved. Most houses are now made from brick or concrete and more than three-quarters have access to electricity. The residents have been given ownership of the land they once lived on illegally and many services like shops and banks have been established within the area. Self-help schemes help boost community spirit as neighbours work together to help each other. However, it relies on the motivation of the local people to succeed.

Figure 7.28: *The Favela Painting project*

Hint

A detailed answer using facts and figures will gain more marks.

Site and service schemes have led to better health and disease prevention by improving access to clean running water and sanitation systems. The Favela Bairro Project has transformed at least 60 of the 600 favelas in Rio. However, a downside is that now the families who live there have to pay rent on their new and improved homes and not all families can afford the rent if they are unemployed. From the government's point of view, a site and service scheme costs roughly two-thirds the price of a Cingapura 'superblock'.

Similarly, favela residents often cannot afford to live in the 'superblock' developments as it costs roughly $60 to initially move in and then between $18 and $26 a month for a 20-year mortgage on one of the flats. The community spirit of the favelas is often lost and people are left feeling isolated. The sites of the new 'superblocks' are often located far from any job opportunities due to the limited building space available.

Since pacification began in 2008, the murder rate and violent crime rates in favelas have decreased. However, innocent bystanders have been killed when police enter favelas and engage in gunfire with the criminal gangs. There have also been widespread reports of police corruption and abuse after the police have taken back control of different areas. This has led to frequent protests and the amenities the residents were promised, such as schools and hospitals, have not yet been built. Many residents find it hard to trust the police, but some describe their neighbourhoods as safer now that there is a police presence.

Need for transport management

Rio de Janeiro is limited in growth potential due to the high mountains that surround the city. These mountains also restrict the movement of traffic within the city, leading to a high volume of vehicles building up on city roads. Public transport is unreliable and trains on the Supervia suburban railway frequently break down. In September 2013, passengers were so enraged by delays that they set fire to several train carriages and vandalised a station.

Figure 7.29: *Train fires in Rio de Janeiro, September 2013*

Much like Glasgow, the subway system in Rio is limited and only covers certain areas of the city. The favelas have no subway coverage, which limits the ability of residents to travel. However, unlike Glasgow, few will have access to their own car due to the cost. The subway, like the trains, also experiences regular breakdowns and is grossly overcrowded. Due to this, many people choose to walk or cycle around the city but the streets are poorly maintained and there is a lack of traffic lights, which makes this option a dangerous one.

The bus network is also heavily overcrowded with more than four million passengers per day. The cost of a single bus journey in Rio de Janeiro is one of the most expensive in Brazil. Heavy traffic on the roads means that it is common for people's daily commute, by bus or by car, to be prolonged. For those on public transport, this means having to stand for long periods of time. Rio geographer Felipe Bagatolli says: 'During the week I spend a lot of time on buses, often standing. I think this affects my quality of life and that of thousands of others, causing health problems because of the fatigue, pain in the legs. And of course, there's the time lost that one would hope to use in other ways, including leisure.'

Figure 7.30: *Traffic in Rio de Janeiro*

Transport management strategies employed

In the run up to the 2014 FIFA World Cup and the 2016 Olympic Games, Rio invested hugely in public transportation improvements. Laudemar Aguiar, head of international relations for the city, said, 'Rio is transforming. By 2018, 70% of Cariocas will have access to mass transportation. In 2009, that number was only 18%. Every day we are building more BRT, more LRT, more metro, connecting the city, and making it better for everyone.'

Figure 7.31: *The Maracana Stadium, the location of the opening and closing ceremonies for the 2016 Olympic Games*

On 1 June 2014, the city opened the second of four bus rapid transit (BRT) systems, called the TransCarioca. The new 39 km exclusive bus lanes encompass 45 stations and five terminal stations, linking the international airport in the north with residential and commercial areas of the city. Each bi-articulated bus is air-conditioned and can carry 180 passengers, thus transporting more people more quickly than traditional bus services. The TransCarioca BRT is well integrated with bus and light rail networks, providing good connectivity across Rio. The new BRT draws 270 000 daily users but this is expected to increase to 320 000 when the system is fully operational.

Figure 7.32: *The TransCarioca BRT*

Rio de Janeiro has continued to expand its network of cycle lanes. The city added 3.3 km of new protected lanes and cycle paths throughout the city in 2014. By 2016, the city had created over 450 km of cycle track. They have also installed over 1300 bike racks in key locations since 2003, creating safer and easier options for bikers.

Figure 7.33: *Bike Rio: a project of sustainable transport*

Rio's subway (metro) system opened in 1979 and has two lines covering 26 miles with 35 stations. An extension was added in 2004 that links the tourist area of Copacabana to Ipanema. Further extensions are proposed and completely new lines are planned. An additional line to Barra da Tijuca and Zona Oeste (west zone) is planned in the future to reduce pressure on road bridges and allow a faster commute for passengers.

Buses remain the main mode of public transportation in Rio. To help ease some of the overcrowding, the government runs the buses more frequently during rush hours. Plans are also in place to manage on-street parking by installing on-street meters.

The favelas of Rio de Janeiro are a labyrinth of steep stairs and alleyways that can be just a few feet wide. There is no room for public transit in this narrow maze. However, the Complexo do Alemão, one of Rio's largest slums, has a possible solution – a 3.5-km-long cable car system. As part of the city-wide pacification policy, 155 eight-seat cable cars travel between six stations built across the favela. For the residents, this has transformed what used to be an hour-and-a-half trudge to a nearby rail station into a 16-minute sky ride.

Figure 7.34: *Cable cars above the Complexo do Alemão*

Success of transport management strategies

Little has been done to significantly reduce the number of cars on the road in Rio. However, the new BRT has helped to remove nearly 500 buses from the streets, reducing congestion and emissions across the city. The BRT provides a safe, reliable and easy commute for millions of people travelling to popular destinations in Rio and people have reported that their journey times have decreased by about 65%. Despite this, the BRT remains heavily overcrowded, with many commuters forced to stand for long journeys.

Figure 7.35: *Overcrowding on BRT*

Make the link

You will learn about strategies to reduce greenhouse gas emissions in the Global climate change chapter.

The additional 14-km subway line between Ipanema and Barra da Tijuca is predicted to increase the subway's capacity by 230 000 passengers per day, enabling more efficient travel around the city.

Bike Rio, a public bicycle sharing system in the city, began in 2011. This project has increased awareness of sustainable travel and allowed people to move more freely around the city. The programme promotes a healthy atmosphere and the use of the most sustainable mode of transit, contributing to Rio's goal of reducing greenhouse gas emissions.

The cable car system cost the city roughly $74 million. Jorge Mario Jáuregui, the architect behind the system, says the project has real and symbolic value: 'real because the connection has been built, and symbolic because it makes the informal city part of the formal city.' However, many residents feel this money would have been better spent in areas where there is still no running water or sewers, and street battles between police and drug gangs continue to kill dozens each year. With the 2014 World Cup and the 2016 Olympic Games thrusting Rio onto the world's stage, many feel that this kind of project was more about improving the appearance of Rio and less about improving conditions for those living in the favelas.

Summary

In this chapter you have learned:

- The need for housing management in a developed and developing world city using Glasgow and Rio de Janeiro as examples
- Strategies used to solve housing problems in these cities
- Impact of these strategies – both benefits and problems
- The need for traffic management in a developed and developing world city using Glasgow and Rio de Janeiro as examples
- Strategies used to manage these problems
- Impact of these management strategies – benefits and problems.

You should have developed skills and be able to:

- Analyse data from tables and charts
- Interpret and compare photographic evidence.

End of chapter questions and activities

Quick questions

Glasgow

1. Discuss the main problems of living in the renovated Gorbals housing area at the beginning of the 1960s.

2. Explain how these problems were solved.

3. Discuss the advantages and disadvantages of the Glasgow Harbour Scheme.

4. Explain why there was a need for traffic management in Glasgow.

5. Discuss the methods used to manage traffic in Glasgow.

6. Evaluate the success of the M74 extension.

Rio de Janeiro

7. Explain why there is a need for housing management in Rio de Janeiro.

8. Explain the main problems of favelas.

9. Discuss the methods used by the city authorities to manage the problems.

10. Evaluate the success of these methods.

11. Discuss why there is a need for traffic management in Rio de Janeiro.

12. Explain the improvements made to public transport.

13. Evaluate the success of these improvements.

Exam-style questions

1. The average growth rate of vehicular traffic in Mumbai was 7% over the past seven years, with 300–350 new vehicles on the road every day.

   For Mumbai, or any named city you have studied in the developing world, **evaluate** the strategies employed to manage traffic congestion.

   8 marks

2. With reference to a developed world city you have studied, **explain** the impact of recent housing changes that have taken place in the inner city.

   8 marks

3. With reference to a developed world city you have studied, **explain**, in detail, why there is a need for housing management.

   8 marks

4. With reference to a developed world city you have studied, **discuss** the reasons why there is a need for transport management.

   8 marks

Activity 1: Mind mapping

In groups, brainstorm the reasons for transport problems in and around Glasgow's CBD. Create a large mind map to illustrate the reasons your group have discussed. For each reason, add at least one method used by Glasgow City Council to manage transport issues in and around the city. For each method, add a comment about how successful it has been. Use the internet to find some pictures to illustrate each management strategy and add these to your mind map.

Activity 2: Annotations

In pairs, draw a picture of a typical favela home and surrounding area in Rio de Janeiro. Annotate your drawing to show what problems these homes and areas have and explain what impact they might have on the people who live there, e.g. open sewers – this can lead to the rapid spreading of diseases such as typhoid, cholera and diarrhoea.

Activity 3: Debate

'The M74 extension will bring only benefits to Glasgow and the West of Scotland.'

Use the internet to research the advantages and disadvantages of the M74 extension. The class will be split into two groups; one will argue in support of the above statement and one will argue against.

Learning Checklist

You have now completed the Urban chapter. Complete a self-evaluation sheet to assess what you have understood. Use traffic lights to help you make up a revision plan to help you improve in the areas you have identified as amber or red.

- Explain the problems facing people living in the city's council housing estates. ⬭ ⬭ ⬭

- Evaluate the problems of living in Glasgow's high-rise flats. ⬭ ⬭ ⬭

- Explain the strategies used to manage these post-war problems including:

  ➤ comprehensive redevelopment areas ⬭ ⬭ ⬭

  ➤ tenement renovation ⬭ ⬭ ⬭

  ➤ new towns. ⬭ ⬭ ⬭

- Explain the successes and failures of these strategies. ⬭ ⬭ ⬭

- Discuss recent strategies to manage Glasgow's housing including:

  ➤ redevelopment of the Gorbals ⬭ ⬭ ⬭

  ➤ Red Road flats ⬭ ⬭ ⬭

  ➤ Glasgow Harbour ⬭ ⬭ ⬭

  ➤ athletes' village. ⬭ ⬭ ⬭

- Evaluate the successes and/or failures of these developments. ⬭ ⬭ ⬭

- Explain the need for traffic management in Glasgow. ⬭ ⬭ ⬭

- Discuss transport management strategies including:

  ➤ roads ⬭ ⬭ ⬭

  ➤ railways ⬭ ⬭ ⬭

  ➤ subway/underground ⬭ ⬭ ⬭

  ➤ pedestrians. ⬭ ⬭ ⬭

- Evaluate the success of these strategies. ⬭ ⬭ ⬭

- Explain the need for housing management in Rio de Janeiro, Brazil. ⬭ ⬭ ⬭

- Discuss issues of shanty towns including:

  ➤ poor quality housing

  ➤ lack of basic services

  ➤ overcrowding

  ➤ disease

  ➤ unemployment

  ➤ crime/gangs

  ➤ landslides.

- Explain methods used to deal with these problems including:

  ➤ self-help schemes

  ➤ site and service schemes

  ➤ Projeta Cingapura (superblock schemes)

  ➤ pacification campaigns

  ➤ non-government agencies/charities.

- Evaluate the success of these strategies.

- Explain the need for traffic management in Rio de Janeiro.

- Assess management strategies used including:

  ➤ public transport

  ➤ buses

  ➤ railways

  ➤ underground

  ➤ cycle lanes.

- Evaluate the success of these strategies.

Glossary

Amenities: facilities within the home such as baths, toilets, hot water, etc.

Basic services: facilities in houses like running water, electricity, sewage disposal.

Bi-articulated bus: another carriage added on to increase the length of a bus allowing it to carry far more passengers.

Brownfield site: urban land that has previously been developed, such as the site of an old factory.

Bus lane: dedicated lane for buses at certain periods throughout the day.

By-pass: road built around a busy urban area to avoid traffic jams.

CBD: Central Business District or city centre; the commercial and business centre of a town or city.

Cholera: water-borne disease prevalent in areas of poor housing.

Commuters: people who live in one area and travel to work in another area.

Congestion: overcrowding on roads causing traffic jams.

Conurbation: a large urban settlement that is the result of towns and cities spreading out and merging together.

Council estates: housing provided by the local council.

Derelict: term used for abandoned buildings and waste land.

Detached house: a house standing alone (not joined to another) typical of the wealthy suburb zone of a city.

Disease: ailment, e.g. cholera, malaria, etc.

Favela: area of slum housing in Brazil.

Grid iron: street pattern running parallel to each other in different directions forming a grid.

Illegal settlements: poor quality housing, like favelas, built on land without planning permission.

Inner city: the part of the urban area surrounding the CBD.

High density: lots of housing/buildings built close together and housing large numbers of people, e.g. tenements or favelas.

High rise: multi-storey flats.

Low-density housing: detached or semi-detached houses with outside space.

Migrant: somebody who moves from one place to another.

Modernise: give something a new/more up-to-date appearance.

Overcrowding: too many people in an area.

Pacification: promotion of peace and putting an end to conflict.

Park and ride: parking of cars at a bus or train station outside town then using public transport to travel into town.

Pedestrianisation: streets where cars are not allowed.

Planned: organised.

Pollution: contamination by smoke, fumes, exhaust emissions, etc.

Regenerate: stimulate and redevelop an area.

Rehouse: move people out of one place to another.

Renovate: improve.

Residential: area where people live.

Ring road: a by-pass that provides a route around the CBD.

Run-down: in bad condition.

Rural: countryside.

Self-help schemes: methods designed to allow people in poor areas to help themselves to survive rather than depend on others.

Semi-detached house: a house joined to one other. These are common in the middle-class suburb zones of a city in the developed world.

Shanty town: an area of poor quality housing, lacking in amenities such as water supply, sewerage and electricity.

Shopping mall: a modern shopping centre.

Site: piece of land something is built on/found.

Slum: a house unfit for human habitation.

Squatters: people living illegally in a house or area.

Tenements: large residential blocks of flats built in the inner cities of developed countries during the industrial revolution to house workers in high-density cramped and unhygienic conditions next to the factories.

Terraced house: a house within a long line of joined housing.

Unemployment: being without a job.

Upgrade: improve.

Urban renewal/regeneration: the improvement of old houses and the addition of amenities in an attempt to bring new life to old inner city areas.

Urban sprawl: the unplanned, uncontrolled growth of urban areas into the surrounding countryside.

Vandalism: damage/defacement of property.

Zone of transition: the inner city area around the CBD. It is a zone of mixed land uses, ranging from car parks and derelict buildings to slums, cafes and older houses, often converted to offices or for industrial use.

Global Issues

8 River basin management

Within the context of River basin management you should know and understand:

- Physical characteristics of a selected river basin
- Need for water management
- Selection and development of sites
- Consequences of water control projects.

You also need to develop the following skills:

- Analyse and synthesise information from a range of numerical and graphical information
- Interpret a wide range of numerical and graphical information.

Introduction

A river is a physical feature of the landscape where water is naturally channelled towards a large body of water like the sea or a lake. A river basin is the land area that is drained by a river and its tributaries. However, humans can alter and control this natural channel to use for their own purposes. When this happens we say that the river is 'managed'. One of the most managed rivers in the world is the Colorado River in the United States of America.

Make the link

You learned about drainage basins in the Hydrosphere chapter.

Hint

In the exam, you may be given a number of resources for a basin (such as a map, climate graphs) and asked to explain the need for water management. Study the resources carefully and include information from them in your answer – they have been provided for a reason!

Colorado River case study

Figure 8.1: *The Colorado River in Arizona*

Figure 8.2: *The Colorado River basin*

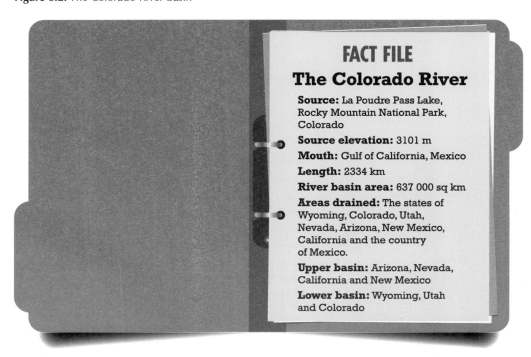

FACT FILE
The Colorado River

Source: La Poudre Pass Lake, Rocky Mountain National Park, Colorado

Source elevation: 3101 m

Mouth: Gulf of California, Mexico

Length: 2334 km

River basin area: 637 000 sq km

Areas drained: The states of Wyoming, Colorado, Utah, Nevada, Arizona, New Mexico, California and the country of Mexico.

Upper basin: Arizona, Nevada, California and New Mexico

Lower basin: Wyoming, Utah and Colorado

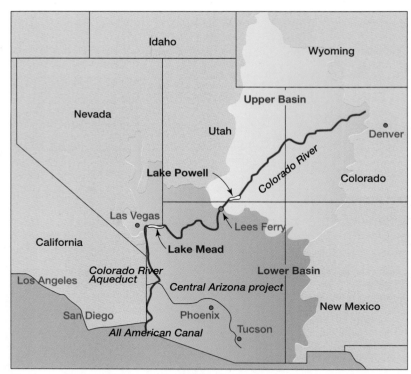

Figure 8.3: *Colorado River: upper and lower basins*

Physical characteristics of the Colorado River basin

The Colorado River rises in the mountains of Colorado and flows in a south-westerly direction for approximately 2334 kilometres until it empties into the Gulf of California in Mexico. It falls 3000 metres along its course. Significant amounts of water are added by tributaries originating in Wyoming, Colorado, Utah, New Mexico, Nevada and Arizona including the Green River in Colorado and the Gila River in Arizona.

The river and its tributaries drain approximately 631 800 square kilometres. The river basin is approximately 1440 kilometres long and 480 kilometres wide in the northern part and 800 kilometres wide in the southern part. The upper portion is one of high elevations and narrow valleys. The lower portion has lower elevations, wide basins and deserts. The main rock types are resistant sandstone and limestone.

A canyon section in northern Arizona and southern Utah creates a division of the Colorado River basin. The Colorado River basin is divided into the upper basin, where waters naturally drain into the Colorado River above Lees Ferry, and the lower basin, where waters drain into the Colorado River below Lees Ferry

> ### 🔍 Hint
>
> Revise your case studies carefully! You will be asked to include real examples from a case you have studied in several of the questions in the exam.

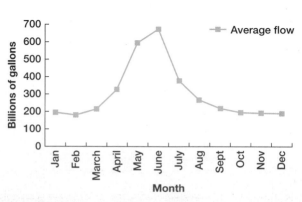

Figure 8.4: *Hydrograph of the Colorado River*

Need for water management

Figure 8.5: *Map of western USA and Colorado River*

Make the link

Hydrographs are explained in the Hydrosphere chapter.

Hint

You should make reference to:

- Climate
- Domestic water supply for increasing population
- Power supply for expanding cities and industry
- Water for irrigation as food demands increase
- Flood control
- Political tension.

Climate

The Colorado River basin has **low annual rainfall** within the lower basin states (Arizona, Nevada, California and New Mexico). In some of these areas there is less than 250 mm of rainfall per year, making these areas desert. Therefore, there is a need for river management to ensure these areas get water. In the upper basin states (Wyoming, Utah and Colorado) the rainfall and snowmelt are extremely variable and unpredictable. Therefore, there is a need to manage the river to ensure these states have enough water from one year to the next. In some parts of the river basin, temperatures reach over 40°C; this makes any water evaporate very quickly so there is a need to manage the river to ensure enough water is available for all of the people who live within the basin.

Domestic water supply

Cities found within the Colorado River basin such as Las Vegas and Phoenix are some of the **fastest growing cities** in the USA. Therefore, they need water to enable their continued development. Water in these areas is needed for swimming pools, flushing toilets and drinking water. As the population in these desert areas continues to increase, the **demand for domestic water supply** continues to rise.

Make the link

You learned about the demand for water in desert areas in the Rural chapter.

Make the link

You will learn more about renewable energies such as hydroelectric power in the Global climate change chapter.

Figure 8.6: *Swimming pools at a hotel in Las Vegas*

Power supply

Cities such as Las Vegas and Phoenix also need more and more **power for industrial and domestic use** as they grow. Hydroelectric power (HEP) is a cheap and clean source of power, which comes with managing the flow of the Colorado River using dams.

Food demands

As more and more people move into areas alongside the river, there is **greater demand for food production** to support this increasing population. However, due to the variable and unpredictable rainfall along the west coast of the USA, this area needs ten times more irrigation water than the east coast to support its population. To enable this production, the Colorado River must be controlled.

Flood control

Prior to the Colorado being dammed it used to flood regularly, creating damage and destruction. By managing the river flow, developments could take place along the river without the **risk of flooding**. Additionally, the river contained a lot of reddish sediment before it was dammed, making the water dirty. Today the dams act as a huge trap for the sediment. Thus, the water is cleaner as it flows downstream.

Figure 8.7: *Residents look over the flooded Colorado in Denver, Colorado in 1983*

Political tensions

Finally, the Colorado flows through seven states and two countries. Each state and country feels it is entitled to the most water, thus fuelling many political arguments. River basin management helps to control the amount of water that each state receives, thus providing a **fairer allocation of the Colorado's water**.

Effect of water management schemes on the hydrological cycle

As discussed in chapter 2, the hydrological cycle is the movement of water around the Earth, from the land to the sky and back again. Humans can interfere with this natural process when they attempt to control and manage river basins.

Evaporation

When rivers like the Colorado are dammed, surface run-off is reduced. Additionally, there is an increase in evaporation from the surface of the massive reservoirs created behind the dam. For example, Lake Mead loses approximately 211 cm of water per year through evaporation. Dams restrict water flow so there is less water flowing below the dam and into the sea. There is also less evaporation from the Colorado River itself as it contains less water due to storage in the reservoirs.

> ## ○ Hint
>
> You should make reference to:
> - Evaporation
> - Storage
> - Infiltration
> - Transpiration
> - The water table
> - Microclimates.

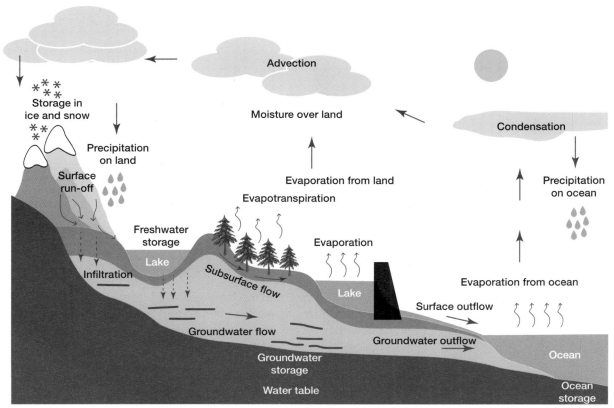

Figure 8.8: *The hydrological cycle*

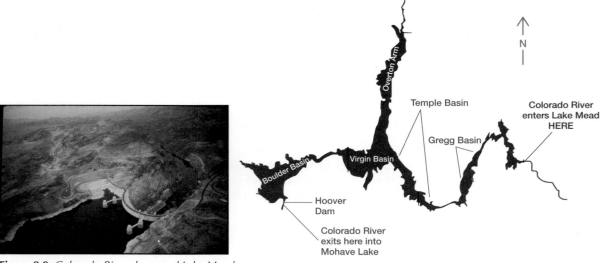

Figure 8.9: *Colorado River dams and Lake Mead*

Infiltration

The amount of water infiltrating the ground is increased as a result of storage in reservoirs. Infiltrating rates into the ground are also altered as rivers are diverted for irrigation channels, such as the Colorado Aqueduct, which takes water to California.

Furthermore, there may be a change in seasonal variations in river levels and the level of **water tables** may also change.

Figure 8.10: *Colorado Aqueduct*

Transpiration

Removing forestry from the dam/reservoir construction sites can lead to less transpiration and less interception. However, if forestry is then planted after a dam has been constructed, it will have the opposite effect, with an increase in transpiration. This increase in moisture in the air will lead to an increase in rainfall, altering the climate of the area. Large lakes can also create their own **microclimates** by keeping surrounding land cooler or even increasing precipitation.

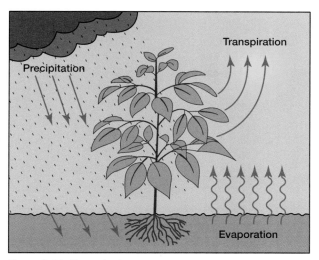

Figure 8.11: *Transpiration*

Industry and housing

Finally, industry and housing will be attracted to the area due to the cheap and clean source of power being generated by HEP. This increases air pollution and can affect global warming, which will alter the hydrological cycle.

Physical and human factors affecting the site of a dam

Choosing the site of a dam depends on many factors. These factors can be split into two categories: physical factors and human factors.

Physical factors

1. Rock type

There must be solid foundations to secure the dam from failure. Although the chief engineer of the Glen Canyon Dam once stated that he would have preferred the harder igneous rock (granite), the canyon walls and the sandstone bedrock can safely support a large 10 million ton dam.

Hint

Read questions carefully as they could be asking for human or physical factors, or both.

Figure 8.12: *Glen Canyon Dam walls*

2. Landforms

Narrow valley: there must be a narrow cross-section to reduce the length of the dam needing to be built and also reduce the strains upon the structure once the lake behind is full. The Colorado has eroded a narrow steep valley about 750 ft (228 m) wide at Glen Canyon and this meant a narrow dam could be built, reducing costs.

Make the link

You may have learned about river erosion for National 5 Geography.

Figure 8.13: *Glen Canyon Dam: narrow cross-section*

Deep valley: a large deep valley must be available to flood behind the dam to store as much water as possible. At Glen Canyon, the canyon is over 700 ft (213 m) deep and therefore an ideal site for a dam and reservoir. The lake created behind it (Lake Powell, below) is 266 square miles (689 square km).

? Did you know?

The lake has a shoreline of 1960 miles (3150 km).

Figure 8.14: *Lake Powell*

3. Climate

There must be sufficient water flow to fill the new lake behind the dam and maintain supplies for all the parts of the scheme. It is important to note local evaporation rates to work out the potential natural losses.

Figure 8.15: *The new lake behind the dam*

4. Permeability of rock

It is important to work out the permeability of the rock below to work out losses through seepage. The sandstone of the Glen Canyon Dam is not totally impermeable, so there is some water loss due to seepage.

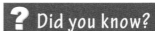
❓ Did you know?

Lake Powell took 17 years to fill, reaching maximum storage in 1980.

Figure 8.16: *Glen Canyon Dam sandstone*

Human factors

1. Loss of farmland

It is important to factor in the amount of farmland to be flooded in order to ascertain the cost payable to landowners and if the loss can be balanced by the additional land irrigated.

Figure 8.17: *Farmland*

Hint

Be able to discuss human and physical factors in your answer.

2. Flooding existing settlements

The number of settlements to be flooded and the likely cost of moving and compensating people and companies from the areas affected also need to be factored in.

3. Local objections

If the dam is built in an area where there is a low population density then there is also likely to be less opposition to the dam's construction.

? Did you know?

Rainbow Bridge, the largest natural bridge on Earth, is at risk from the rising waters of Lake Powell.

Figure 8.18: *Local objectors*

4. Local services and amenities

However, the further away the dam is from centres of high population, the further the distance to transport electricity, or water for irrigation to customers. It would need to be determined if the distance was cost efficient. Greater distances also mean greater losses from evaporation and seepage.

Social, economic and environmental advantages and disadvantages of water management projects

For any river basin management control project, there are a number of advantages and disadvantages. These can be split into categories: social, economic and environmental.

> ## 🔍 Hint
>
> Read the question carefully. It could be advantages or disadvantages and refer to either social, economic or environmental.

Social advantages

- Improved water supply for drinking as the dams ensure a constant supply.
- Irrigation water allows farmers, for example in California, to grow crops all year round. This increases food supply.
- The availability of water can sustain increasing populations especially in the desert cities of Phoenix and Las Vegas. The Colorado River provides water for over 40 million people.
- There is greater availability of electricity.
- The local populations around Lake Mead reservoir now have access to recreational activities like water sports, available because of the creation of the reservoir for the Hoover Dam.
- The Hoover Dam is a tourist attraction, bringing money into the area. The facilities built for the tourists improve the social life of the local populations.

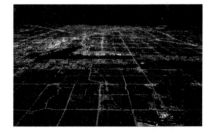

Figures 8.19–22: *Social advantages*

> ## ⚛ Make the link
>
> If you study Physics, you will learn more about the generation of electricity.

Social disadvantages

- Many people were forced to leave their homes to allow the flooding of the valleys to create the reservoirs. The town of St Thomas was drowned beneath the waters that created Lake Mead. The people had to be resettled in other areas, which were often not as productive as the land they left.

Figures 8.23–24: *Social disadvantages*

Economic advantages

- The dams produce cheap HEP, attracting industry into the area, providing jobs and increasing the standard of living.
- The water and power produced has encouraged the expansion of the cities of Las Vegas and Phoenix.
- The facilities in these cities attract many tourists, bringing money into the economy.
- The availability of irrigation water for farming means agricultural produce increases. This extra produce for sale creates more income for the economy.

Make the link

You may have learned about tourism as part of National 5 Geography.

Figures 8.25–28: *Economic advantages*

Economic disadvantages

- The schemes cost a huge amount of money to build and maintain.

- The more irrigation water is used by farmers along the river, the more saline the water becomes.

- At first, when the water eventually reached Mexico it was unusable. To solve this problem, a huge desalinisation plant had to be built at Yuma, costing over $300m to build, with running costs of around $20m a year.

- Water is very cheap for farmers to buy, so much is wasted. The cost of delivering water to farmers is around $350 per acre foot but only costs farmers $43.50 per acre foot. Hence, they do not try to conserve it, resulting in a huge loss to the economy.

- Large amounts of compensation need to be paid out to people and farmers to be relocated as the reservoirs flood their land.

Make the link

If you take Biology, you will learn more about farming and the production of food today.

Figures 8.29–31: *Economic disadvantages*

Environmental advantages

- There is a constant supply of water for domestic use, which benefits health.

- The creation of reservoirs encourages wildlife into the area. More than 250 species of birds have been counted in the Lake Mead area.

- The reservoirs and dams are seen by some people as improving the scenery and environment.

Figures 8.32–34: *Environmental advantages*

Make the link

If you take Biology, you will learn more about the importance of biodiversity.

Environmental disadvantages

- The original wildlife in the area have been forced to move as their habitats have been destroyed by the reservoirs, etc. There are no longer wild beavers in Tucson.
- The natural landscape has been destroyed by the development of the dams and associated reservoirs.
- The level of Lake Powell is so high that the Rainbow Bridge, said to be one of the geological wonders of the world, is slowly being dissolved by the water.

> ## 🔍 Hint
> Remember, although bullet points like those used here are useful for your notes, in an exam, questions should always be answered in full sentences.

Figures 8.35–36: *Environmental disadvantages*

Political problems of a river running through more than one state or country

River basins can cross states or international boundaries, causing difficulties in cooperation between states and countries. For example, the Colorado River runs through seven states of the USA, and Mexico. This can cause many political problems and arguments as to how to divide up the river's waters.

Different states have different laws on water rights. The amount of water allowed to reach the lower areas depends on the cooperation of upstream neighbours. The quality of the water reaching the lower areas again depends on the water use further up the river. The areas lower down the valley can be left with very little water, which can be polluted due to industrial use further upstream.

The waters can also be used for irrigation and repeated withdrawal of the river water leads to salinity downstream. In some cases the water quality is so poor it needs to be treated, e.g. by desalinisation plants like the one in Yuma. These are very expensive and arguments occur as to which state should pay for them.

Colorado River Basin

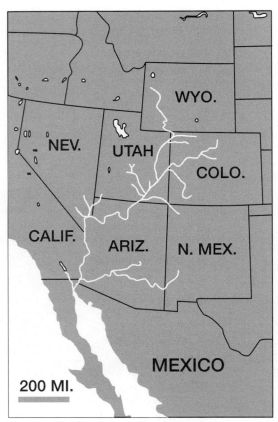

Figure 8.37: *Map showing the Colorado River crossing state and country boundaries*

All parties have to agree to the allocation of the Colorado water. Different states have different needs for domestic use, agriculture and industry. Each state can pass laws regarding water but these can conflict with each other so it needs the federal government to take control. Each state must agree to contribute to the cost and maintenance of the dams and reservoirs.

Figure 8.38: *Workers at the desalination plan in Yuma*

The legal system in the USA worked in favour of the richest state (California), so it had more power over the water than the other states, creating conflict. Mexico was unhappy because, as the last area to receive the water, the water quality was poor. Agreements were needed to establish who paid for the cleaning of the water, e.g. by the desalinisation plant at Yuma. By the time the water reaches Mexico, there may not be enough water left to meet its allocation.

> ## Make the link
>
> If you study the USA in Modern Studies, you will learn more about its federal government and legal system.

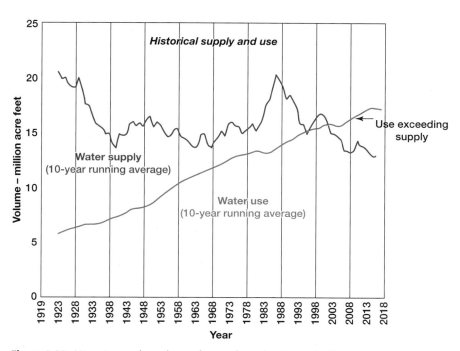

Figure 8.39: *Historic supply and use of water from the Colorado River*

The future?

In 2013, the American Rivers Society named the Colorado River as the most endangered river in the USA. The Colorado River is endangered partly due to outdated and wasteful water management that is inadequate to meet the modern needs of the states, as well as the persistent drought of recent years. Too much water is being drained from the river, resulting in problems for communities and the environment. The Colorado River is a lifeline in the desert areas of the basin. Its water sustains over 40 million people in seven states, as well as in Mexico. It irrigates over four million acres of farmland, which grows over 15% of the USA's crops. However, demand on the river's water now exceeds its supply, leaving it so overused that it no longer flows to the sea. New legislation is now needed to ensure water supply sustainability in the Colorado River basin.

Figure 8.40: *The river is so overused that it no longer flows to the sea*

Summary

In this chapter you have learned:

- General reasons why rivers need to be managed
- The physical characteristics of the Colorado River
- Detailed reasons why the Colorado needs to be managed
- Human factors considered in the siting of a dam
- Physical factors considered in the siting of a dam
- The social advantages and disadvantages of managing the Colorado
- The economic advantages and disadvantages of managing the Colorado
- The environmental advantages and disadvantages of managing the Colorado
- The political advantages and disadvantages of managing the Colorado
- Future problems of managing the Colorado.

You should have developed skills and be able to:

- Identify the location and states of the Colorado basin from a map
- Interpret a hydrograph of the Colorado River
- Analyse graphs on river flow over a period of time
- Compare graphs on supply and demand of the Colorado's water.

End of chapter questions and activities

Quick questions

1. Discuss the physical features of the Colorado basin.

2. Explain the physical and human factors that need to be considered when selecting and developing sites for dams.

3. Explain the benefits of your chosen river basin management scheme. You should refer to social, economic and environmental consequences. Exchange your answer with your partner. Mark it and make at least one positive comment and one suggestion as to how it could be improved.

4. Explain the problems of your chosen river basin management scheme. You should refer to social, economic and environmental consequences.

5. Discuss why extracting water from the Colorado River could cause problems for the basin states and Mexico in the future.

Exam-style questions

1. Study Figures 8.41–44.

   a. **Explain** the need for water management in the Irrawaddy River in Myanmar.

   10 marks

   b. For a named river basin management scheme you have studied, **explain** the physical factors that should be considered when selecting the site for the dam and its associated reservoir.

   10 marks

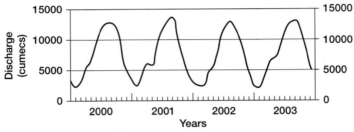

Figure 8.41: *Monthly discharge of the Irrawaddy River at Myitsore before the HEP scheme*

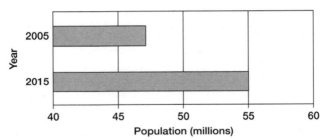

Figure 8.42: *Population change in Myanmar*

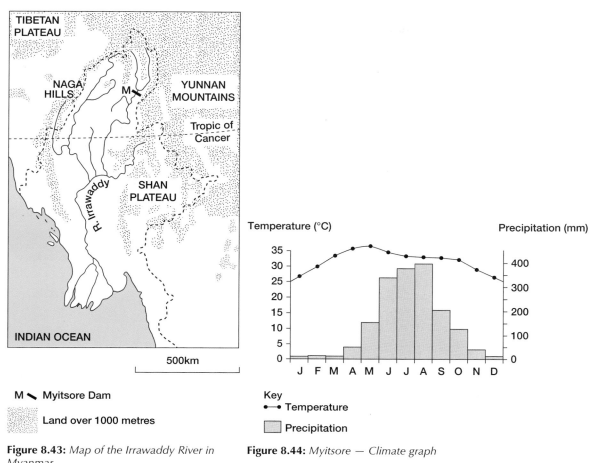

M ↘ Myitsore Dam

Land over 1000 metres

Key
•—• Temperature

Precipitation

Figure 8.43: *Map of the Irrawaddy River in Myanmar*

Figure 8.44: *Myitsore — Climate graph*

2. **Discuss** the social, economic and environmental benefits of a named water management project you have studied.

    10 marks

3. **Discuss** the social, economic and environmental adverse consequences of a named water management project you have studied.

    10 marks

Activity 1: Thought shower

a. Discuss with a partner or in a group the reasons for the need to control a river. Create a spider diagram showing your results. Share your results with the rest of the class, adding any new reasons to your diagram.

b. With your partner or group, discuss the ways in which a river can be 'managed' to meet these needs. Now add these to your spider diagram.

Activity 2: Information leaflet

Look at the text on pages 181–183 on the need for river basin management.

With a partner or in a group, produce an information leaflet on the benefits of water management schemes.

Activity 3: Debate

Debate – 'Should rivers be managed?'

Divide into groups. Prepare a speech to support your choice. Debate your position with the rest of the class.

Learning Checklist

You have now completed the River basin management chapter. Complete a self-evaluation sheet to assess what you have understood. Use traffic lights to help you make up a revision plan to help you improve in the areas you have identified as amber or red.

- Demonstrate a knowledge of the reasons why a river needs to be managed.

- Identify the Colorado River and the basin states from a map.

- Describe the physical features of the Colorado basin.

- Describe the climate of the Colorado basin.

- Evaluate the problems caused by the climate of the Colorado basin.

- Explain the human reasons why the Colorado needs to be controlled.

- Discuss the physical factors that need to be taken into account when selecting a site for a dam.

- Discuss the human factors that need to be taken into account when selecting a site for a dam.

- Explain the social advantages and disadvantages of controlling the Colorado River (be able to give named examples).

- Explain economic advantages and disadvantages of controlling the Colorado River (be able to give named examples).

- Explain the environmental advantages and disadvantages of controlling the Colorado River (be able to give named examples).

- Explain the political advantages and disadvantages of controlling the Colorado River (be able to give named examples).

- Understand the future problems of continuing to remove water from the Colorado.

- Understand a hydrograph/river flow graph.

Glossary

Aqueduct: a channel or pipe to carry water over huge distances.

Communications: roads, railways and air travel.

Compensation: payment given to a person to make up for personal loss, e.g. land flooded by a reservoir.

Desalination plant: the industrial removal of salt from water.

Drought: lack of rainfall over a period of time.

Evaporation: water lost from bodies of water like lakes.

Flooding: the submerging of land under water.

Habitat: area where plants and animals live.

HEP (hydroelectric power): energy produced from fast-flowing water.

Irrigation: artificially watering crops using sprinklers or irrigation channels.

Permeability: the ability of a liquid to pass through a rock.

Reservoir: a place where water is collected and stored for future use.

River management: controlling a river.

Seasonal rainfall: rainfall occurring at a specific time of the year.

Sediment: deposits like sand and pebbles carried along by a river.

Sustainability: healthy and endurable for the future.

Tributary: small stream joining a larger one.

Unpredictable: erratic.

Variable: changeable.

9 Development and health

Within the context of Development and health you should know and understand:

- Validity of development indicators
- Differences in levels of development between developing countries
- A water related disease: causes, impact, management
- Primary health care strategies.

You also need to develop the following skills:

- Interpret a wide range of numerical and graphical information
- Analyse and synthesise information from a range of numerical and graphical information.

Introduction

Development refers to the social and economic status of a country. In other words, whether a country is socially and economically 'rich' or 'poor'. Any improvement in the standard of living of people is called development.

The northern half of the world can be described in general as the **developed world**. All of the countries found here are comparatively **rich** and most of their citizens enjoy a high quality of life. The southern half of the world can be described as the **developing world**. The quality of life in most countries found here is, on the whole, much lower. Many people living in these areas do not have access to life's essentials of food, clothing, shelter, water, medical care and education.

> **? Did you know?**
> Around 80–85% of the world's population currently live in developing world countries.

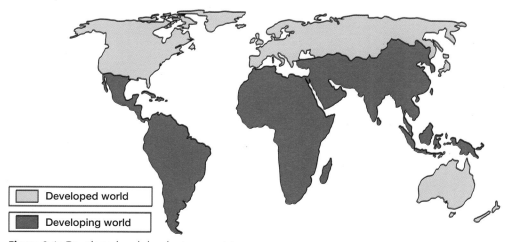

| | Developed world |
| | Developing world |

Figure 9.1: *Developed and developing countries*

Indicators of development

There are a number of ways in which to measure development. You can use: social indicators, economic indicators or combined indicators.

Social indicators show how a country uses its wealth to improve the lives of its people. Economic indicators measure the wealth and industrialisation of a country. A table displaying some social and economic indicators is shown below:

	Social indicators						Economic indicators			
	Population (millions) (2012)	Life expectancy at birth (2011)	Crude death rate per 1000 people (2011)	Birth rate per 1000 people (2011)	Infant mortality rate per 1000 live births (2012)	Adult literacy rate (%) (2011)	Energy consumption per capita (kWh) (2011)	Gross national income per capita ($) (2012)	Gross domestic product per capita ($) (2012)	Motor vehicles per 1000 people (2010)
Brazil	198.7	73	6	15	13	90	2438	11630	11339	209
China	1350.7	75	7	12	12	95	3298	5720	6091	58
India	1236.7	66	8	21	44	63	684	1580	1489	18
Kenya	43.2	60	9	36	49	72	155	860	942	24
Malawi	15.9	54	12	40	46	61	...	320	268	8
Nepal	27.5	67	7	22	34	59	106	700	690	5
Sierra Leone	6.0	45	18	38	117	44	...	580	634	6
Spain	46.2	82	8	10	4	98	5598	29620	28624	593
United Kingdom	63.2	81	9	13	4	99	5516	38670	39093	519
United States	313.9	78	8	13	6	99	13246	52340	51748	797

... No data available

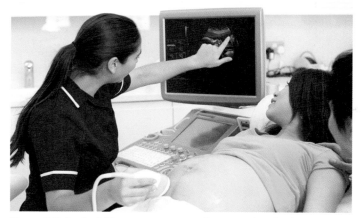

Figure 9.2: *In the UK, pregnant women have regular checkups*

One social indicator is **infant mortality rate**. This is the number of deaths of infants under one year of age per 1000 live births in any given year. In developed countries such as the UK, this figure is about 4/1000. This low figure shows that the level of health care within the country is good. The majority of babies are born in well-equipped hospitals and are given vaccinations in the first year of life to protect them against killer illnesses. In poorer, developing countries, such as Sierra Leone, infant mortality rates are much higher at 117/1000. This shows that health care, particularly in rural areas, is not good and many children die as a result of problems at birth or from diseases, such as cholera and malaria. There is also a lack of clean water with many children suffering from starvation and malnutrition, which makes them weak and more prone to illness.

Gross Domestic Product (GDP) is an example of an economic indicator. It can show how wealthy a country is as it measures the total value of goods produced in a country in US dollars per capita. If GDP is low, it suggests very little secondary or tertiary industry. In Malawi, it was about $505 in 2019. Most of the people work in agriculture, most as subsistence farmers, only producing enough food to feed themselves without contributing anything to the economy. In the UK, the GDP is much higher, at approximately $43 200. This indicates that many people are employed in industry or service industries that bring wealth into the country. Countries such as the UK also make money from trade.

> ### ⚫⁘ Make the link
>
> If you have studied Business Management, you will have learned about the sectors of industry.

Figure 9.3: *A market in Malawi*

How reliable are social and economic indicators?

The simple answer is that they are not very reliable if taken on their own. Levels of development must be based on a number of indicators used together rather than individual indicators on their own such as Gross National Income (GNI). The use of single indicators can be misleading and often fails to illustrate the actual picture of life in an individual country for a number of different reasons:

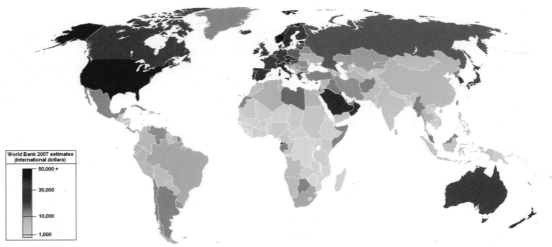

Figure 9.4: *Gross National Income per capita, 2007*

> ### Hint
>
> Remember to mention specific examples in your answer. Detail increases marks.

1. Indicators are **averages** and they can hide huge differences that exist in every country. For example, even in developed countries like the UK, differences exist between different areas of the country. For example, there are differences between the north and the south of England – standards of living are generally higher in the south of England where the capital city of London can be found. In Brazil, there are huge differences in the standard of living between the south-east and Amazonia in the north.

2. Indicators in developing countries can also hide any **rural/urban differences** that exist. For example, in Brazil people in rural areas generally live in poorer conditions than those living in urban areas. They may not have the same access to doctors, education or clean water.

3. **GNI can also be inflated** in countries that have oil. Countries like Kuwait and Saudi Arabia are high-income countries but since the wealth is held by only a small number of people, standards of living for the majority are much lower.

4. You cannot trust the figures in many developing countries because **census data may be inaccurate**. For example, China stretches over a wide area and takes in huge tracts of desert and mountain ranges. Therefore, poor communication routes mean some places are inaccessible. Also, there are many different languages spoken, so many people do not understand the census and give inaccurate answers.

Composite indicators

It is generally accepted that individual indicators are of limited value, for all the reasons previously mentioned. A country with an above average calorific intake indicator may have well-fed people, but this does not tell us much about their overall health, education, wealth or social freedom. Therefore, to provide a more accurate view of a country as a whole it is possible to group indicators together. These are called composite or combined indicators.

There are two main composite indicators: the Human Development Index (HDI) and the Physical Quality of Life Index (PQLI).

Human Development Index

HDI is a combination of five social and economic indicators: life expectancy, cost of living, adult literacy rate, school enrolment and GNI per capita. This produces a number between 0 and 1. An HDI of 0.8 or above is considered developed.

Physical Quality of Life Index

This is a combination of three social factors: life expectancy, infant mortality rate and adult literacy rate. This gives a number between 0 and 100. A PQLI of over 77 is considered developed.

Taking all of these measures together gives a more accurate insight into levels of development and makes it easier to compare countries.

Figure 9.5: *Some areas of China are inaccesible, making it difficult to collect data*

Make the link

You learned about the problems with census data in the Population chapter.

Differences in levels of development between developing countries

Factor affecting development	Example country/area	Reason for level of development
Resources	Saudi Arabia Kuwait	Although Saudi Arabia is still considered a developing country, it has access to raw materials such as oil which are in demand globally. This allows them to trade with other countries and brings in money which could be spent on improving the country. However, the money is not always spent on the people and is controlled by a small percentage of the population who enjoy a high standard of living while the majority of people remain poor.

Industry	South Korea Malawi	South Korea is a Newly Industrialised Country (NIC) and has developed, as it offers educated, resourceful and cheap labour to foreign companies who invest there. This brings money into the country which has been invested in education, health care and developing the infrastructure of the country. Countries such as Malawi have a high percentage of people employed in subsistence agriculture and this does not bring money into the country. Malawi is also a landlocked country, which makes it harder to develop good trade links.
Climate	The Sahel region of Africa, e.g. Chad/Ethiopia	The Sahel struggles to develop as it is based on a nomadic farming culture with land degradation. This area experiences recurring drought and, therefore, has associated problems, such as famine, to contend with. Development is hindered due to a lack of money.
Health	Kenya Ghana Uganda	If a country has an endemic disease like malaria, people will be too ill to work. This means that the economy will suffer and the country will struggle to develop. The country will be caught in the 'vicious circle of poverty'.

(continued)

Politics	Afghanistan Somalia Zimbabwe	The Afghan government spends a large majority of the country's money on military equipment, which means that there is less money available for the development of housing, industry, health care and education. Civil wars and political instability can also lead to disruption and other countries will not invest or trade with countries such as Somalia and Zimbabwe.
Economy/Debt	Ethiopia Chad Mali	Some countries suffer from drought. This makes it very difficult to grow food to feed the population. The country gets into debt as the government has to borrow money to help the people instead of investing in projects such as education and health care.
Relief	Nepal	Countries with difficult relief, like Nepal, find it difficult to develop. This is because it is problematic to build communication links or industries. Countries like this are unlikely to attract foreign investment.

Natural disasters	Bangladesh	Countries that suffer from natural disasters struggle to develop. This is because the government spends so much money on disaster relief that it has no money left for development. Bangladesh suffers from monsoon rains and flooding. It also struggles to develop its tourism as people are not attracted by this kind of climate so it struggles to make money to develop.
Population	Uganda	A high population growth will generally limit development, because resources will increasingly have to be spread more thinly (e.g. food, space, jobs and water). There will not be enough jobs, houses, schools, health clinics, etc., and as a result development will be hindered.
Tourism	Thailand	Countries can actively encourage tourism, which can bring in foreign currency and provide job opportunities. This money can then be used to develop the country in other aspects, such as health care and education.

Malaria: causes, impact and management

Although a number of species can transmit the disease, the best-known species is the female anopheles mosquito, which bites an infected person, sucking blood containing the malarial parasite. The mosquito hosts this parasite and then bites another person, transmitting the parasite into a new victim. We can attempt to reduce the impact of malaria by avoidance, prevention and control, and by treatment.

Figure 9.6: *A boy with mosquito bites*

Malaria usually occurs in the tropics and affects over two billion people across 100 countries. Over 400 million people in Africa alone live in areas where there is malaria.

Make the link

You may have learned about wealth inequality, civil war and political instability in Modern Studies.

You have looked at issues of drought and water management in the Hydrosphere, Rural and River basin management chapters, and at the issues caused by large and growing populations in the Population chapter.

Make the link

You may have studied natural disasters and tourism in detail for National 5 Geography.

Hint

Compare this map to Figure 9.1 at the beginning of this chapter – what do you notice?

Malaria Risk Areas

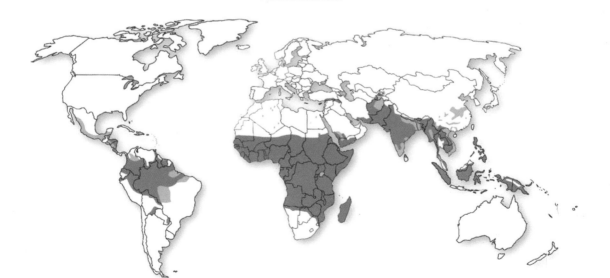

☐ No malaria
▨ Areas with limited risk of malaria transmission
▨ Areas where malaria transmission occurs

Figure 9.7: *Location of malaria around the world*

Physical factors affecting the location of malaria	Human factors affecting the location of malaria
Temperatures of 16–40°C, with high humidity.	Mosquitoes require a 'human blood reservoir' – many people to feed from.
Mosquitoes are usually found in areas with an altitude of up to 3000 metres.	Populations are increasing and people are able to move around more easily – this may lead to the disease being introduced to areas where it is not currently prevalent.
The mosquito requires areas of shade in which to digest the blood meal.	In areas where marshlands have been drained and/or dams, reservoirs or irrigation ditches built, there will be pools of stagnant water in which mosquitoes can breed. This is also true of places like Vietnam where there are bomb craters from a previous war.
The mosquito lays its larvae in stagnant pools of water such as lakes, paddy fields and puddles.	The spread of malaria is encouraged by bad sanitation and poor drainage – these things tend to be a problem in poorer areas.

Make the link

You have looked at the management of water in the River basin management chapter, and migration in the Population chapter.

Figure 9.8: *Paddy fields are a breeding ground for mosquitoes*

Strategies used to control malaria

Strategy	Effectiveness
Education programmes These are run from local villages. People are educated on the disease and are taught the benefits of covering up arms and legs to prevent the mosquitoes biting. Try to reduce levels of poverty as stronger, healthier people can help fight off malaria better.	Education is effective if delivered via song and play in areas where literacy rates are low.
Villages >1 km away Keep villages and people at least 1 km away from mosquito breeding grounds.	This is effective because people are more at risk the closer they live to mosquitoes, so if they live further away they are less likely to contract malaria.

DDT Insecticide sprayed on ponds, walls, trees, etc.	This was effective but eventually the mosquitoes became immune. It was environmentally damaging. It was expensive so developing countries could not afford it. Newer insecticides, such as malathion, have been developed. As this is petrol-based, it is less harmful to the environment, but homes and breeding grounds have to be sprayed more frequently and this makes it expensive to do. It also stains the walls of people's houses a horrible yellow colour and villagers complain about the smell, so some don't want it sprayed in their homes.
Chloroquine This is a drug that people take to protect them from contracting malaria. Also used to treat malaria.	This was effective but soon became less effective as the parasite became able to overcome it. Drugs are also expensive, which means that developing countries struggle to afford them. Often, people suffer side effects from the drugs.
Spray oil or egg white on breeding grounds This puts a layer of oil or egg white on top of the stagnant water, which drowns and suffocates the mosquito larvae.	Although this kills the mosquitoes, the oil can damage the environment. Oil and eggs are also too expensive for developing countries to use. Food is scarce in many developing countries, so using egg whites can be seen as a waste of a valuable food resource.
Infect coconuts with Bti bacteria Infecting coconuts with this bacteria and then throwing them into stagnant water will encourage the larvae to eat them.	This is effective because when the larvae eat the coconuts the bacteria kills them.
Sprinkle mustard seeds Sprinkle mustard seeds on stagnant water.	This is effective as the mustard seeds are sticky and the larvae will stick to the seeds and sink down into the stagnant water, which will drown them.
Fish By putting fish into areas of stagnant water, mosquito larvae will be eaten by muddy loach fish.	This stops the mosquito population from multiplying. This is effective as the fish can also be eaten by the humans to add protein to their diet.
Drainage of breeding grounds The areas where mosquitoes breed, like ponds, swamps and paddy fields are drained.	This was effective because mosquitoes cannot survive without stagnant water. Unfortunately it is impossible to drain EVERY breeding ground. In tropical areas there can be heavy rainfall every day, so new areas of stagnant water can appear all the time.
Building dams The water trapped behind the dam can be released; this then flushes out the streams and breeding grounds, which drowns the larvae.	Although this is effective, a lot of human labour is required, which makes it impractical and expensive.
Bed nets These are placed over beds at night to stop mosquitoes from biting people.	These are very effective in stopping the spread of malaria as they are cheap (about £5) and easy to introduce into communities.

(continued)

Cover windows and doors with gauze Gauze works like bed nets as it stops mosquitoes from entering people's homes. Mosquitoes like to bite ankles or arms, especially in darkened rooms. Gauze stops mosquitoes from getting in.	This is an effective method as it is cheap.
Planting eucalyptus trees These trees soak up excess water, which helps to drain breeding grounds.	This is a reasonably effective method as it reduces stagnant water and also reduces the breeding grounds of the female anopheles mosquito.
Genetic engineering This method created a large pool of sterile male mosquitoes. This would eventually reduce the entire mosquito population. Also, only the FEMALE mosquito carries the parasite for malaria.	This eventually proved too expensive. Many people think that genetic engineering is morally wrong.

Make the link

You looked at the building of dams in detail in the River basin management topic.

If you take Biology, you will learn more about genetic engineering.

Hint

If you are asked to comment on the effectiveness of a strategy, make sure you include detail – say why it is or is not effective; there may be several reasons. If there are differing viewpoints on its effectiveness, make sure you include these and say why the different groups disagree.

Figures 9.9–11: *Some strategies for combatting malaria include education, planting eucalyptus trees, and the use of bed nets*

How malaria can hinder development

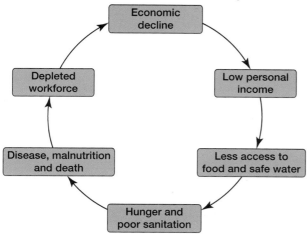

Figure 9.12: *Cycle of poverty*

Treating the disease is costly. Governments can spend almost 40% of their budgets on fighting malaria, e.g. in Malawi. That money could be spent on other things, e.g. building schools and hospitals, improving housing and sanitation, providing clean drinking water and reducing the occurrence of other diseases such as cholera.

The country loses productivity as people are often too ill and weak to work. If malaria were controlled, then productivity would increase, more food could be grown, people would be better fed, more children would survive so infant mortality rates would fall. Families would be smaller as there would be less need to have lots of children and more children would go to school, obtain a good education and get more skilled jobs. This all constitutes greater development. Surpluses could be sold, generating trade.

Tourism might increase and multinational corporations would be more likely to create bases in a country if there was less threat of catching malaria. This would create thousands of jobs and would generate much needed money for the country, helping to improve people's standard of living.

Primary health care

Primary health care (PHC) strategies have been introduced by many developing countries in an effort to improve the health of the population. Their governments do not have the money to provide the sophisticated health care systems found in richer countries like the UK so they decided to put in place basic provisions to try to supply health care. These strategies are called primary health care. The health care is on a simple level, providing the people with information on nutrition and diet, contraception, supplying commonly needed drugs, e.g. anti-malarial drugs and the treatment of minor medical problems. Some strategies used are outlined overleaf.

Figure 9.13: *A barefoot doctor visits a patient in China*

Barefoot doctors are local people who are given basic training so they can look after their communities' health needs. They treat common or simple ailments, from snake bites to appendicitis. This means that medical help is available within easy reach of the community and saves people from travelling to the nearest hospital. It takes pressure off the large hospitals, allowing them to deal with more serious cases. Training costs are low. In India it cost around $100 to train a health worker for a year. They provide families with advice on birth control, vaccinations and basic hygiene. This helps to prevent the spread of disease, and to reduce birth rates and infant mortality rates. They also run clinics in the larger villages and take a mobile health van to more remote villages. The advantage of this is that they are well known and trusted by the villagers to provide them with vital health care. If the cases are too complicated, they are referred to the nearest hospital where highly trained doctors and nurses can provide a wider range of facilities and medical care. PHC programmes also try to provide villages with local dispensaries. This gives the people access to essential modern drugs, traditional remedies and family planning.

Use of **ORT (oral rehydration therapy)** to tackle rehydration. This is an easy, cheap and effective remedy for diarrhoea and dehydration and is particularly used for babies.

The provision of **vaccination programmes** against disease such as polio, measles and cholera. It is cheaper to try to prevent the disease than treat it when it is caught. Prevention is better than cure.

Figure 9.14: *A community health worker vaccinating a child against polio in Afghanistan*

Talks on health education are given in schools and touring groups put on plays and sing songs about health problems to give people information on issues like AIDS. Many people in the developing world are illiterate, so this type of oral education is more suitable than producing leaflets and other forms of written instruction and advice.

Figure 9.15: *Schoolgirls learn about AIDS by playing a board game in Johannesburg, South Africa*

Improved sewage/effluent disposal (including toilets) and small-scale **clean water supplies** can be built – often with the participation of members of the community. An example of this is solar water disinfection (Sodis). This is a type of portable water purification that uses solar energy. This is effective as it is cheap and makes biologically contaminated water safe to drink.

Small local health centres staffed by generalist doctors (similar to GPs) may be built. The doctors here can conduct minor procedures. However, this is more expensive than the other PHC strategies and may require funding from other countries.

Figure 9.16: *A man learns about water management in Burkina Faso, Africa*

Problems with primary health care

Despite the success of barefoot doctors, there are not enough of them to care for everybody and some lack the medical skills required, which can lead to inappropriate prescribing of drugs and incomplete surgery.

In addition to this, the system often breaks down when funding is not available for further training, or is withdrawn or reduced.

Summary

In this chapter you have learned:

- The definition of a developed and developing country
- Ways of identifying the level of development of a country
- Descriptions of a variety of indicators of development
- How reliable social and economic indicators are
- The make-up and use of composite indicators like HDI and PQLI
- The advantages and disadvantages of composite indicators
- Reasons for differences in development between developing countries
- How malaria is transmitted
- The physical factors affecting the spread of malaria
- The human factors affecting the spread of malaria
- How malaria can hinder development
- Strategies used to control malaria
- How effective these strategies are
- The definition of primary health care
- Methods used to promote primary health care
- Impact of primary health care.

You should have developed skills and be able to:

- Analyse development indicators from a table
- Identify countries/areas from a map affected by a disease
- Compare the level of development of an area from a map
- Interpret the effects of a disease from diagrams.

End of chapter questions and activities

Quick questions

1. Name one social indicator and one economic indicator of development and discuss how they show a country's level of development.

2. Explain the advantages of using a composite indicator of development such as the HDI rather than a single indicator.

3. Referring to named developing countries that you have studied, account for the wide range in levels of development between developing countries.

4. Explain, in detail, the physical and human causes of malaria.

5. Explain the benefits to a developing country of controlling the disease.

6. Discuss methods used to try to control the spread of malaria.

7. Evaluate the effectiveness of these methods.

8. Evaluate the effectiveness of some specific primary health care strategies employed in developing countries you have studied.

Exam-style questions

1. Infant mortality rate per 1000 live births is a social indicator of development.

    Name one other social indicator and one economic indicator of development and **explain** how they show a country's level of development.

    6 marks

2. Referring to named developing countries that you have studied, **suggest reasons** why there is such a wide range in levels of development between developing countries.

    10 marks

3. Study Figures 9.17 and 9.18.

Many African countries have been trying to eliminate water-related diseases like malaria, cholera and bilharzia/schistosomiasis.

For one of the above diseases:

a. Discuss the physical and human factors that put people at risk of contracting the disease.

10 marks

b. Evaluate the measures that can be taken to combat the disease.

8 marks

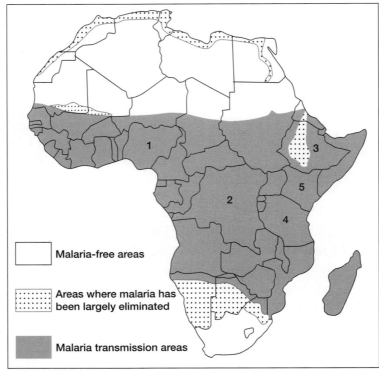

Most infected countries

1 Nigeria (57.5 million)
2 DR Congo (23.6 million)
3 Ethiopia (12.4 million)
4 Tanzania (11.5 million)
5 Kenya (11.3 million)

In brackets are the number of people infected.

Malaria-free areas

Areas where malaria has been largely eliminated

Malaria transmission areas

Figure 9.17: *Map of areas in Africa affected by malaria*

- 3.3 billion people in 109 countries are at risk from malaria
- 247 million annual cases of malaria
- 850 000 people die from malaria each year
- 91% of deaths caused by malaria are in Africa
- 85% of deaths caused by malaria are of children aged under five

Figure 9.18: *Statistics on malaria*

4. Study Figure 9.19.

   Many developing countries are attempting to reduce the death rate of children under five by implementing primary health care strategies.

   Discuss some specific primary health care strategies and explain why these strategies are suited to people living in developing countries.

   12 marks

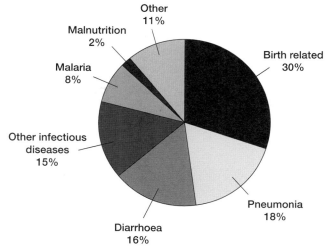

Figure 9.19: *Major causes of death in children under five in the developing world, 2008*

Activity 1: Internet/ newspaper research

Over the next week you are asked to look for articles on primary health care in newspapers or research primary health care on the internet. Print the articles or cut out the articles from the newspaper. Create a summary of the facts discussed in each article. Create a collage of the articles your class found and display your facts about primary health care around the edges of the collage.

Activity 2: Create a blog

Individually write a blog appealing for help to control a disease like malaria. You should include:

- a world map showing the areas affected
- a description of how malaria is transmitted
- a description of the effects of malaria
- an explanation of what can be done to combat malaria.

Include tables, graphs or statistics in your blog.

Learning Checklist

You have now completed the Development and health chapter. Complete a self-evaluation sheet to assess what you have understood. Use traffic lights to help you make up a revision plan to help you improve in the areas you have identified as amber or red.

- Demonstrate an understanding of the difference between a developed and developing country.
- Explain the meaning of indicators of development.
- Explain and give examples of a social indicator of development.
- Explain and give examples of an economic indicator of development.
- Explain a composite indicator.
- Describe, in detail, the PQLI.
- Describe, in detail, the HDI.
- Evaluate the advantages of using a composite indicator.
- Explain the reasons for the differences in development between developing countries. Using named examples, you should be able to discuss resources, industry, climate, health, population, tourism, relief, natural disasters, politics, economy and debt.
- Describe how the mosquito spreads malaria.
- Describe the location of affected areas from a map.
- Discuss the physical factors affecting the spread of malaria including temperature, altitude, shade and water.
- Discuss the human factors affecting the spread of malaria including poverty, population growth, stagnant water, great movement of people.
- Explain strategies used to control malaria, e.g. education programmes, bed nets, fish, genetic engineering, draining swamps.
- Evaluate the strategies used to control malaria.
- Explain how malaria can affect the development of a country.
- Explain primary health care.
- Discuss some primary health care strategies, e.g. you should be familiar with barefoot doctors, education programmes, ORT, vaccination programmes, improvement in hygiene.
- Discuss the suitability of primary health care.

Glossary

AIDS: Acquired Immune Deficiency Syndrome – a serious disease of the immune system that is caused by infection with a virus.

Bed nets: preventative measure to reduce malaria.

Birth rate: the number of babies born per 1000 of the population.

Chloroquinine: drug used to prevent and treat malaria.

Composite indicator: a group of indicators used to show the development of a country.

DDT: used as a pesticide but banned in many countries for its adverse effect on the environment.

Death rate: the number of deaths per 1000 of the population.

Debt: money owed to another country.

Developed country: countries with a good standard of living.

Developing country: countries with a poor standard of living.

Development indicator: a measurement of a country's development.

Drought: lack of rainfall over a period of time.

Effluent: liquid waste or sewage discharged into a river or the sea.

Endemic: a disease that is constantly present in some degree in people of a certain class or in people living in a particular area.

Family planning: methods used to control the size of a family.

Genetic engineering: the deliberate modification of the characteristics of an organism by manipulating its genetic material, e.g. sterile male mosquitoes.

HDI: Human Development Index – a combined indicator which measures a country's development based on social and economic indicators.

Hygiene: conditions or practices (such as cleanliness) conducive to health.

Infant mortality rate: the number of children who die before the age of one.

Life expectancy: the number of years a person is expected to live.

Malaria: a disease spread by the anopheles mosquito.

Malnutrition: the condition that develops when the body does not get the right amount of the vitamins, minerals and other nutrients it needs to keep healthy.

Oxfam: charity working towards reducing poverty.

PQLI: Physical Quality of Life Index – a combined indicator used to measure a country's development.

Prevalence: the number of cases of a particular disease present in a population at a given time.

Prevention of disease: methods to avoid occurrence of a disease.

Primary health care: provision of basic health care.

Rehydrate: the process of replacing lost fluids to the body through loss from diarrhoea, etc.

Stagnant water: standing water where mosquitoes can breed, encouraging the spread of disease.

Starvation: condition caused by lack of food.

Strategy: method or scheme.

Transmit: spread a disease from one person to the next.

Vaccination: inoculation to prevent a specific disease.

10 Global climate change

Within the context of Global climate change you should know and understand:

- Physical and human causes
- Local and global effects
- Management strategies and their limitations.

You also need to develop the following skills:

- Interpret a wide range of numerical and graphical information.

What is global climate change?

'We have to face the reality of climate change. It is arguably the biggest threat we are facing today.'

William Hague

Former British Foreign Secretary

Make the link

You learned about heat exchange in the Atmosphere chapter.

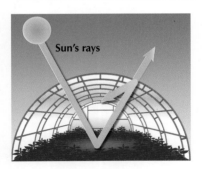

Figure 10.1: *The greenhouse effect*

All around the Earth are gases. These gases trap heat from the Sun and warm up the Earth's surface so that life can exist here. Some gases act a bit like a greenhouse – they allow the Sun's rays to pass through but they also prevent them from escaping easily as well. This is known as the 'greenhouse effect'. Only about 1% of the gases in our atmosphere are greenhouse gases but without them our planet would be a very cold place with an average temperature of –18°C! If there are too many of these 'greenhouse gasses' in the atmosphere, they can end up trapping too much heat. This is often called 'global warming'.

There are six greenhouse gases but the last four are of particular interest:

1. Water vapour (H_2O)
2. Ozone (O_3)
3. Methane (CH_4)
4. Nitrogen dioxide (NO_2)
5. Carbon dioxide (CO_2)
6. Sulfur dioxide (SO_2)

Make the link

You may have learned more about these gases in Chemistry.

Scientists have gathered a lot of evidence that our climate is changing – there have always been major climatic changes throughout the Earth's history. The graph below shows the last 130 years in detail.

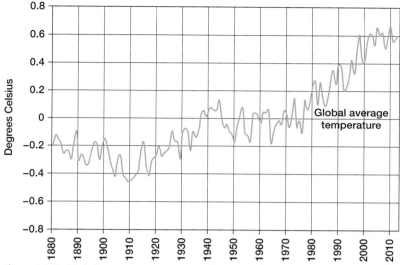

Figure 10.2: *Global average temperature: departure from 1951–1980 average*

What are the main causes of global climate change?

Greenhouse gases are essential to life on Earth. Without them we would freeze, but too many of them and we will overheat. There are many ways in which excess greenhouse gases can be emitted into our atmosphere and begin the process of global climate change. These changes can be split into two categories: physical (natural) causes and human (man-made) causes.

Physical causes of global climate change

1. **Changes in the tilt of the Earth towards or away from the Sun** can affect the amount of insolation that it receives. The greater the tilt of the Earth towards the Sun, the closer some areas are to the Sun and the greater the amount of energy received in these areas, e.g. the Earth's polar regions.

2. Similarly, the Earth rotates around the Sun in an **elliptical orbit** and, therefore, at certain times of the year the Earth can be much closer to the Sun and so receive greater amounts of the Sun's energy. This is called the 'Milankovitch theory'.

> **Make the link**
>
> You learned about the seasonal tilt of the Earth in the Atmosphere chapter.

3. Additionally, cycles of **Sunspot activity** have been linked to an increase in global temperature. The output from the Sun is not constant. Cycles have been identified where there is either a rise in the amount of solar energy or a decrease. Sunspots are dark spots that appear on the Sun. If there are lots of them it means more solar energy is being sent out from the Sun towards the Earth.

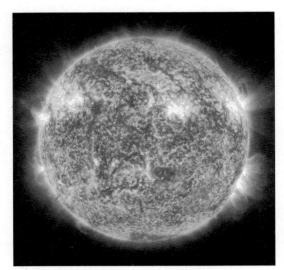

Figure 10.3: *Sunspot activity*

4. **Volcanic eruptions** can release huge amounts of ash, as well as greenhouse gasses, such as sulfur dioxide, into the atmosphere. The sulfur dioxide combines with water in the atmosphere to form sulfuric acid droplets called aerosols. These droplets absorb radiation from the Sun and therefore heat themselves and the surrounding atmosphere. This also means that the Earth's surface does not receive as much insolation so global temperatures may drop after a large volcanic eruption, like the eruption of Krakatoa in 1883.

Make the link

You may have studied volcanoes in detail as part of National 5 Geography.

Figure 10.4: *Volcanic erruptions contribute to climate change*

5. **Oceanic circulation and continental drift** can, over a long period of time, cause periodic warming (El Niño) and cooling (La Niña) of tropical areas in the Pacific Ocean, which can affect the climate of these areas and lead to natural disasters such as heavy flooding.

Figure 10.5: *Floods in Thailand in 2011 were thought to have been caused by the effects of El Niño and La Niña*

Human causes of global climate change

1. There has been a huge **increase in fossil fuel consumption** since the industrial revolution began about 250 years ago. Factories, power stations, increasing numbers of cars on the roads and heating homes has led to the burning of carbon-based fuels, such as coal, oil and natural gas. These fossil fuels release the greenhouse gas CO_2 into the atmosphere. This increases the 'greenhouse effect' and causes temperatures to rise as a result.

Hint

Read the question carefully. It can be physical or human causes or both.

Figure 10.6: *Cars release greenhouse gases into the atmosphere*

Make the link

You learned about the effects of deforestation in the Rural chapter.

Figure 10.7: *Landfill sites emit large amounts of methane into the atmosphere*

Make the link

If you take Biology, you will study modern farming methods.

? Did you know?

A cow releases on average between 70 and 120 kg of methane per year.

2. **Deforestation** in the tropical rainforests can further contribute to the level of CO_2 in the atmosphere as CO_2 is released when the cleared trees are burned. Additionally, there is less absorption of CO_2 as there are fewer trees around the world to absorb the gas, which further leads to a build-up in the atmosphere.

3. **Methane** is increasing in our atmosphere due to an increase in cattle herding, rice production and landfill sites.

   a. **Cows** are some of the greatest methane emitters. Their grassy diet and multiple stomachs cause them to produce methane, which they then exhale with every breath. The huge numbers of cow herds, bred for human consumption around the world, make a significant contribution to global warming.

   b. Paddy fields account for around 20% of human-related emissions of methane. In places like India and China, rice is a staple food so farmers grow large quantities of it in flooded **paddy fields** throughout the growing season. This means that methane is produced by microbes underwater as they help to decay any flooded organic matter.

   c. Humans produce a huge amount of waste and most of this ends up in **landfill sites**. Due to waste slowly decaying and breaking down underground, methane can be emitted for many years, even after a landfill site has been closed.

4. Chlorofluorocarbons (CFCs) are greenhouse gases made up of chlorine, fluorine and carbon. These compounds are increasing in our atmosphere due to increasing human use of **aerosols, fridges and air conditioning units**.

5. **War** (bombs, atomic testing, etc.) creates a large amount of dust which is thrown up into the atmosphere. Much like ash from volcanic eruptions, this dust can reduce the insolation received at the Earth's surface, thus reducing temperatures.

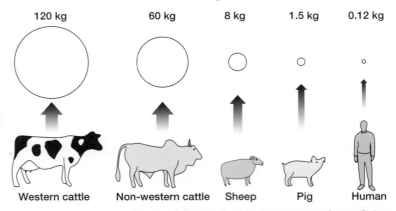

SOURCE: Nasa's Goddard Institute for Space Science

Figure 10.8: *Methane emissions per animal/human per year*

Figure 10.9: *Dust created by bombings contributes to the greenhouse effect*

How is our climate changing? – local and global effects

'We know our climate is already changing. We see it all around – record rainfall and flooding, droughts and wildfires, and worrying reports of shrinking ice shelves breaking up in Greenland. The Intergovernmental Panel on Climate Change has made clear that global average sea levels may rise by as much as 0.83 m by 2100 – placing ever greater pressure on our coastal heritage and communities affected by coastal flooding.'

Paul Wheelhouse MSP

Former Minister for Environment and Climate Change

Local effects of climate change in Scotland

We can already see evidence that Scotland's climate is changing. In the last 50 years, scientists have recorded a rise in sea levels, changes in our rainfall patterns, an increase in extreme weather events and an overall warming of our climate. The last 10 years have been the warmest since records began and the last three decades have seen rainfall greatly increase across Scotland.

According to the Scottish government, as laid out in 'Climate Ready Scotland: Scottish Climate Change Adaptation Programme' (May 2014), the following are some of the consequences faced by Scotland due to climate change.

Make the link

If you take Biology, you will learn about the importance of biodiversity.

You have looked at soils, water and coasts in detail in the Biosphere, River basin management and Lithosphere chapters.

Agriculture and forests	Pests and disease
A warming climate has the potential to improve growing conditions in Scotland and increase the productivity of our agriculture and forestry. Plant growth needs temperatures on average above 6°C so a rise in temperature means plants can grow for a longer period of time.	As our climate changes, the new conditions will allow existing pests and disease to spread and new threats to become established in Scotland. This will impact on the health of our people, animals, plants and ecosystems if risks are not properly managed.
Soils	**Natural environment**
We rely on soils to sustain biodiversity, support agriculture and forestry, regulate the water cycle and store carbon. Soils and vegetation are being altered by changes to rainfall patterns and increased temperatures.	Climate change will affect the delicate balance of Scotland's ecosystems. Some distinctive Scottish species will struggle and could be lost, invasive non-native species may thrive, whilst a degraded environment may not be able to sustain productive land or water supply.
Water	**Flooding**
As our climate warms and rainfall patterns change, there will be increased competition for water. Summer droughts will become more frequent and more severe, causing problems for water quality and supply in Scotland.	With climate change likely to alter rainfall patterns and bring more heavy downpours, flood risk will increase in the future. This will impact on properties and infrastructure – with serious consequences for our people, heritage, businesses and communities.
Coasts	**People**
Sea level rise is already having a widespread impact on parts of Scotland's coast. With this set to increase over the coming decades, we can expect to see more coastal flooding, erosion and coastline retreat – with consequences for our coastal communities and supporting infrastructure.	A warming climate could provide more opportunity to be outdoors and enjoy a healthy and active lifestyle. However, increasing temperatures could adversely affect people, with an increase in skin cancers.
Cultural heritage and identity	**Energy supply**
The changing climate is already altering our unique Scottish landscape and threatening our historic environment through coastal erosion, flooding and wetter, warmer conditions. Also, rising temperatures could reduce the number of snow days, affecting the ski industry.	Climate change will influence Scotland's capacity to generate weather-dependent renewable energy. For example, varying water availability will affect hydro generation schemes.

Figure 10.10: *A forest at Loch Ness*

Figure 10.11: *The coast at Oban*

Figure 10.12: *Skiing in the Cairngorms*

Global effects of climate change

However, the effects of climate change are not only limited to Scotland. Many of the impacts discussed above are impacts that will be felt in many countries around the world. Unfairly, many of the countries impacted most by climate change are the poorer nations that have a very small carbon footprint and contribute little to the problem. There are global consequences that must also be considered.

Make the link

You have looked at issues of farming, water and people in detail in the Rural, Development and health and Population chapters.

Weather	**Sea levels**
Changes in weather are already having an impact around the world. Some places are becoming drier and others are becoming wetter. Some areas are becoming warmer and some areas are becoming cooler. We are likely to see an increasing number of storms, floods and droughts, but we cannot predict which areas of the world will be affected by these. For example, areas around the Mediterranean sea could become deserts, whilst areas like Jersey could suffer from flooding. As these changes take place, plants, animals and even people may find it difficult to survive in the places where they live now.	Higher temperatures will result in ice melting in places like Antarctica and Greenland. This will flow into the sea and sea levels all over the world may begin to rise. Scientists predict that sea levels may rise by as much as 20 to 40 cm by the beginning of the next century. Flooding will become an issue for millions of people in countries like Bangladesh, India and the Netherlands as low-lying coastal areas are at increased risk as sea levels rise. Flooding will become a huge danger and many people will lose their homes or have their livelihoods destroyed as large areas of farmland are ruined.
Farming	**Water**
Just like in Scotland, weather changes will affect the kind of crops that can be grown in places around the world. Crops like wheat and rice grow better in higher temperatures, but other plants, like maize, can't cope with high temperatures. Changes in the amount of rainfall will also affect how many plants grow. As a result of this, some countries, like Brazil, parts of Africa such as the Sahel, Southeast Asia and China, may not be able to grow enough food for their citizens and so people may begin to starve.	Places like the Sahel in Africa already lack enough water for the people living there. Some areas, like Scotland, are likely to experience drier conditions but other areas will experience more rain.
Plants and animals	**People**
Plants and animals around the world will be affected by changes in temperature and rainfall. For example, we are already seeing temperatures rising and polar ice beginning to melt. This will have an effect on the habitats and hunting grounds of polar bears. Polar bears may be forced from their usual habitat into more populated areas. Many plants and animals are at risk of extinction if they are not able to cope with the changes in weather.	Already heavily populated areas may become even more overcrowded as people move inland to avoid the risk of coastal flooding. Countries like China and Egypt that have large coastal populations are most at risk. Diseases like malaria could spread into areas not presently affected as the increase in temperature would allow the mosquitoes to breed.

Figure 10.13: *Amsterdam is a low-lying area that will be affected by rising sea levels*

Figure 10.14: *Maize crops do not grow well in high temperatures*

Figure 10.15: *Polar bears may be forced from their usual habitat as the polar ice melts*

The future

'Climate impacts are already affecting agriculture and food security, human health, water supplies, and ecosystems on land and sea. The world is not prepared yet. Climate change is the single greatest threat to a sustainable future.'

Ban Ki-moon

Former UN Secretary-General

So how can we manage global climate change?

Management strategies and limitations

On an individual level there are many simple strategies that we can all do that will help reduce human causes of global climate change. Making a few small changes can not only reduce greenhouse gases but it can also save you money! Some simple ways that you can reduce your greenhouse gas emissions are shown below:

Switch to energy saving products

Energy saving products, like energy saving light bulbs, can help reduce our role in global climate change. Replace your old lights bulbs with energy saving light bulbs and you will not only help the environment but replacing just one bulb can save you around £3 per year. According to the UK Energy Saving Trust, if you replaced every bulb in your house with an energy saving bulb you could save on average £55 per year! These energy saving bulbs generate less heat, use much less energy and last anywhere from 10–50 times longer than regular bulbs. Make sure that whenever you leave a room, you switch the light off as this will also help save energy.

Figure 10.16: *An energy saving lightbulb*

Seal and insulate your home

The Scottish government and local authorities provide grants to enable people to install loft and cavity wall insulation. This can also save you up to about 20% on your heating bill every year as it prevents leaks by sealing your home and blocking out heat and cold.

Reduce, reuse, recycle

Reducing, reusing and recycling waste at home can help save energy and reduce pollution and greenhouse gas emissions from manufacturing products and disposing of them. Most local authorities in Scotland have a recycling programme in place to recycle things like your newspapers, beverage containers, paper and other goods. South Lanarkshire Council, for example, offers its residents the opportunity to recycle glass, paper, textiles, oil, car batteries and cans at a selection of waste and recycling centres found throughout the area. Composting food and garden waste also reduces the amount of rubbish that is sent to landfills and therefore it reduces greenhouse gas emissions.

> ### 🔍 Hint
>
> Be able to explain how individuals as well as governments can contribute to the reduction of global warming.

Use water efficiently

The water that comes from our tap when we run a bath or have a shower has been pumped into our homes, treated with chemicals to make sure it is clean and safe and heated to make it comfortable to wash with – this takes a lot of energy. We can reduce greenhouse gases by saving water whenever possible; for example, don't let the water run while you are brushing your teeth, have leaky taps repaired right away and only run your dishwasher when it is completely full.

Figure 10.17: *A recycling point in Edinburgh*

Use green power

You can modify your house to generate your own environmentally friendly electricity, e.g. having solar panels installed on your roof or having a wind turbine in your garden. Green power can lower greenhouse gas emissions, and it helps increase clean energy supply.

Figure 10.18: *Solar panels*

Spread the word

Educate your family and friends on the benefits of being energy efficient – it is good for their homes and good for the environment because it lowers greenhouse gas emissions and air pollution.

The management strategies mentioned above can all be very successful if *everyone* adopts them. However, not everyone is willing to make the changes discussed and it can be expensive to do things like install solar panels in your home. Small changes from all of us can make a big difference, though. Why not start today and tell your friends and family how they can help prevent global climate change?

On a larger scale, governments can also agree to work together to reduce emissions of carbon dioxide and other greenhouse gasses. The 2005 **Kyoto Protocol** is an example of a treaty which countries could sign up to in an agreement to reduce their greenhouse gas emissions by 2012.

The Kyoto Protocol introduced a 'carbon credits' scheme. This was basically a way for countries to work together to reduce their emissions. If a country produces CO_2 then it must have a way of containing that CO_2; for example, planting more trees to absorb it. If a country produces more CO_2 than it is able to offset, for example, by planting trees, then it must purchase a 'carbon credit' from another country whose responsibility it then becomes to offset the excess CO_2 according to the Kyoto agreement. The money the country gains from the selling of a carbon credit must be used to promote carbon saving, e.g. promoting wind power or 'fuel switching'.

The Kyoto Protocol had its successes and failures. Many countries, including Scotland, were successful in reducing their emissions by 2012. However, some countries, such as the USA, refused to sign the agreement in the first place so they were not legally bound to meet the targets set by the Kyoto Protocol. Most people agree that there has not been enough progress in preventing global climate change and that the Kyoto targets were not strict enough. As new data becomes available, it is now thought that to prevent significant global climate change, greenhouse gas emissions will need to be reduced by at least 80% around the world as soon as possible!

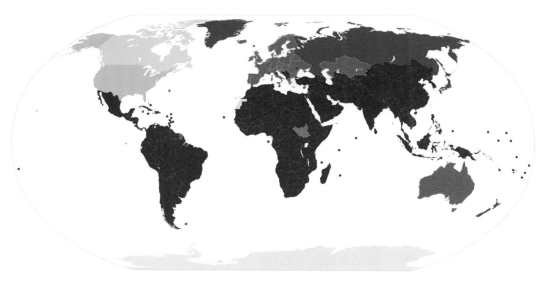

■ Annex B parties with binding targets in the second period	▢ Annex B parties with binding targets in the first period but which withdrew from the Protocol
■ Annex B parties with binding targets in the first period but not the second	▨ Signatories to the Protocol that have not ratified
■ Non-Annex B parties without binding targets	▨ Other UN member states and observers that are not party to the Protocol

Figure 10.19: *Countries ratifying the Kyoto Protocol*

Figure 10.20: *Activists campaign for the USA to sign the Kyoto agreement in 2005*

During the 2011 UN Climate Conference in Durban, many countries agreed to a second commitment period of the Kyoto Protocol. In 2012, a United Nations conference took place on climate change in Doha. After two weeks of talks, 194 nations agreed to implement a second phase of the Kyoto Protocol from 2013 till 2020. The summit established for the first time that rich nations should move towards compensating poor nations for losses due to climate change. It is the only legally binding plan for combatting global warming.

Figure 10.21: *United Nations Climate Change Conference, Doha*

In 2016, the Paris Agreement was introduced. The Paris Agreement is a pact sponsored by the United Nations between 197 countries that focuses widely on reducing greenhouse gases emissions, adapting to the impacts of climate change, and to provide financial assistance to developing countries affected by a changing climate. The Paris Agreement's central aim is to strengthen the global response to the threat of climate change by keeping a global temperature rise this century well below 2 degrees Celsius above pre-industrial levels and to pursue efforts to limit the temperature increase even further to 1.5 degrees Celsius

As people have become more aware of climate change, there has been an increase in the amount of action being taken. During 2019, large numbers of school pupils stayed away from school to march for urgent action on climate change. People are being encouraged to holiday at home to reduce the huge carbon footprint created by air travel. People are also being encouraged to have meat-free days to reduce the harmful methane gases emitted by animals like cows.

Summary

In this chapter you have learned:

- The meaning of global climate change
- The physical and human causes of global climate change
- Local effects of climate change in Scotland
- Global effects of climate change
- Managing climate change on an individual level
- Government attempts to reduce global warming
- The successes/failures of agreements such as the Kyoto agreement.

You should have developed skills and be able to:

- Interpret climate trends from a graph
- Extract data from a pictograph
- Identify countries from a map.

End of chapter questions and activities

Quick questions

1. Explain the 'greenhouse effect'.

2. Discuss how the greenhouse effect contributes to an increase in global temperatures.

3. Study Figure 10.2 on page 225. Describe the variation in global average temperatures from 1880–2010.

4. Discuss the main causes of global warming.

5. Explain why some countries in the world contribute more to global warming than others.

6. Discuss the effects climate change could have in Scotland.

7. Explain the impact of these effects.

8. The effects of climate change are not limited to Scotland. Explain the ways in which climate change can affect the world as a whole.

9. Look at Figure 10.22. Describe, in detail, the extent to which countries are at risk of flooding caused by climate change.

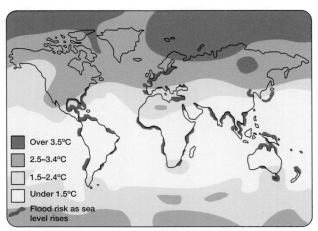

Figure 10.22: *Areas at risk of flooding from rises in sea level*

10. Explain ways in which people can try to reduce global warming.

11. 'Climate change is a global problem. If we're to fix it we need a global solution and we need it soon.' – quote from the World Wildlife Fund for Nature (WWF).

    What role can a) businesses and b) governments play in reducing climate change? How successful are these methods likely to be?

Exam-style questions

1. There has been an increase in the average global temperature in the last 150 years.

   Explain the human causes of global warming.

   8 marks

2. **Explain** the possible consequences of global warming throughout the world.

   10 marks

Activity 1: Global footprint

Draw two footprints onto plain paper.

Divide your footprint into sections. Use keywords in each section to show the main human causes of global warming. Using keywords, label the other footprint to show solutions to these causes.

Use these keywords as the basis for an answer on the human causes and solutions to global warming.

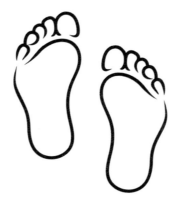

Activity 2: Global warming crossword

You are each going to create your own crossword based on what you have learned about global warming.

Before you start, you should make a note of at least 12 words associated with global warming on a piece of paper (maximum 20).

Beside each word create a clue. Make sure the clue could easily be understood by someone else in the class.

After each clue, put the number of letters in brackets, e.g. (6).

If the answer contains more than one word, show this in brackets, e.g. (5, 4).

You should now access www.crosswordpuzzlegames.com/create.html

In the 'Word' column enter the words you have chosen (see above). If the answer consists of more than one word, do not leave a space between each word, i.e. run all the letters together. For example, global warming should be globalwarming.

In the 'Hint' column you should enter the clue you have made up. (Remember to put the number of letters in brackets.)

When you have completed this, click on 'Create My Crossword Puzzle'.

Now print this off and swap with another class member and try to complete each other's puzzles.

Activity 3: Class debate

In groups you are going to discuss the statement 'Human activity is the main cause of global warming'.

Allocate a role to each group member. Choose from: industrialist, scientist, environmentalist or a student like yourself. Discuss the issue, then make a copy of the diagram below to summarise your opinions.

Be prepared to share your opinions with the rest of the class.

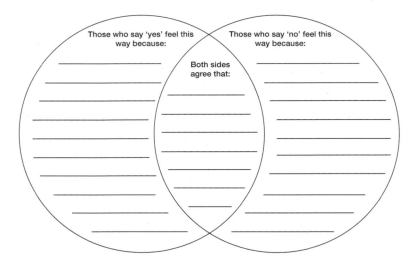

Learning Checklist

You have now completed the Global climate change chapter. Complete a self-evaluation sheet to assess what you have understood. Use traffic lights to help you make up a revision plan to help you improve in the areas you have identified as amber or red.

- Understand the contribution that greenhouse gases make to global warming. ⬭ ⬭ ⬭

- Discuss the evidence from climate graphs used by scientists to show global warming. ⬭ ⬭ ⬭

- Discuss the main human causes of climate change. ⬭ ⬭ ⬭

- Discuss the physical causes of climate change. ⬭ ⬭ ⬭

- Evaluate whether human causes, physical causes or both are the main cause of global warming. ⬭ ⬭ ⬭

- Describe/explain the consequences of global warming on Scotland. You should be familiar with the effects on people and the environment. ⬭ ⬭ ⬭

- Describe/explain the global consequences of global warming on people and the environment. ⬭ ⬭ ⬭

- Discuss methods used to manage climate change. ⬭ ⬭ ⬭

- Explain strategies people can use to reduce global warming. ⬭ ⬭ ⬭

- Explain strategies governments can use to reduce global warming. ⬭ ⬭ ⬭

- Understand how successful these methods are in reducing global warming. ⬭ ⬭ ⬭

Glossary

Aerosols: small particles or liquid droplets in the atmosphere that can absorb or reflect sunlight.

Afforestation: planting new forests on lands that previously have not contained forests.

Alternative energy: energy derived from non-traditional sources (e.g., compressed natural gas, solar, hydroelectric, wind).

Atmosphere: the gaseous area surrounding the Earth.

Biofuels: gas or liquid fuel made from plant material called biomass.

Carbon dioxide: a naturally occurring gas as well as a by-product of human activities like burning the rainforest.

Carbon footprint: the total amount of greenhouse gases that are emitted into the atmosphere each year by a person, family, building, organisation or company.

CFCs: the short name for chlorofluorocarbons – gases that contribute to ozone depletion.

Climate: the average weather conditions over at least 35 years.

Deforestation: the removal of trees from areas like the Amazon rainforest.

Desertification: the changing of land into desert.

Emissions: the release of a substance into the atmosphere.

Fossil fuel: organic materials formed from decayed plants and animals that have been converted to crude oil, coal and natural gas by exposure to heat and pressure in the Earth's crust over hundreds of millions of years.

Global warming: increase in average global temperatures near the Earth's surface.

Greenhouse effect: trapping and build-up of heat in the atmosphere near the Earth's surface.

Greenhouse gas: any gas that absorbs radiation in the atmosphere.

Landfill: land waste disposal site in which waste is generally spread in thin layers, compacted and covered with a fresh layer of soil each day.

Kyoto Protocol: a United Nations treaty which sets legally binding commitments on greenhouse gas emissions.

Ozone layer: the layer of ozone that shields the Earth from harmful ultraviolet radiation from the Sun.

Recycling: collecting and reprocessing a resource so it can be used again.

Renewable energy: energy resources that will not run out.

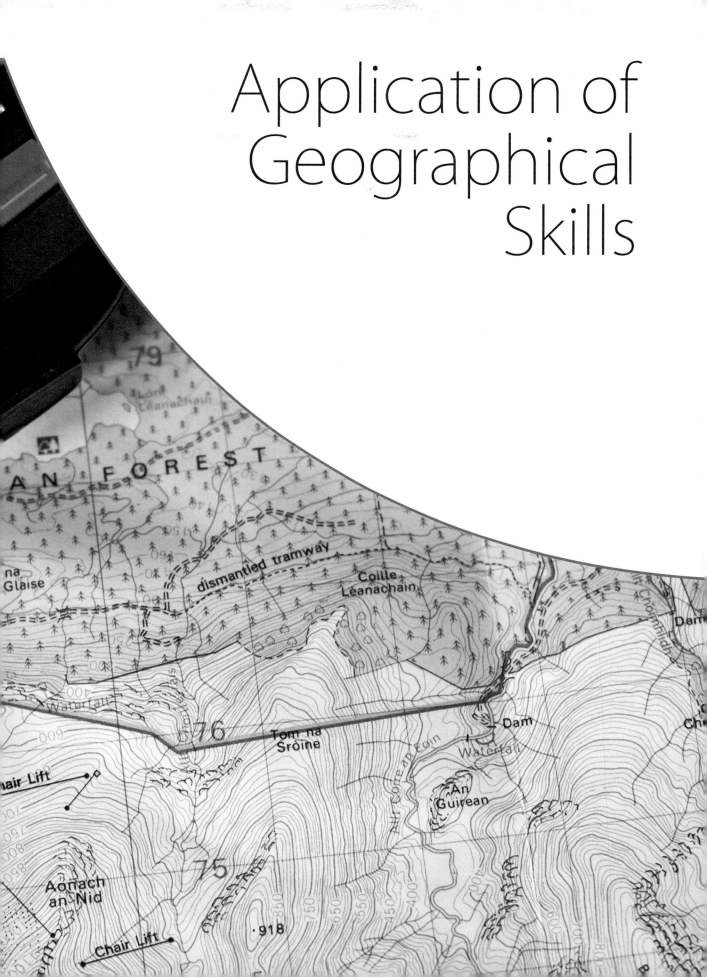

Application of Geographical Skills

11 Application of geographical skills

Scenario question

This is a scenario question and is worth 20 marks. It is compulsory. In this type of question you will be given an Ordnance Survey map along with several other pieces of information, such as a table, chart, map, or newspaper headline. You will then be asked to evaluate and/or assess a particular situation; for example, the suitability of an area for the development of an HEP (hydroelectric power) station or the siting of a quarry. In this type of question you will be able to use the geographical skills you have learned and developed throughout the course.

You will be expected to use the resources given to you to answer the question. You will need to draw conclusions, put forward alternative suggestions and back up your response with map evidence. At this level, when using map evidence you should be using six-figure grid references where possible. Diagrams and charts are there to give you information on the scenario and should be referred to in your answer.

Example question 1

Energy giant Scottish and Southern Energy (SSE) has revealed ambitious plans to build Scotland's biggest hydroelectric scheme. The scheme would require the construction of a new dam and upper reservoir at Coire Glas. It is envisaged that the construction period would last up to five or six years. Working to the conditions below, a site for the dam and reservoir has been proposed.

Conditions for location of a dam

The site should:

- experience sufficient rainfall
- have a narrow, deep valley
- minimise flooding of surrounding farmland and/or settlements
- cause minimal visual impact in the scenic Great Glen area
- cause minimal disruption to local people and settlements.

Study Figures 11.1–6.

Referring to map evidence and other information from the sources, **evaluate the suitability of the proposed site** in relation to the conditions for the location of the dam.

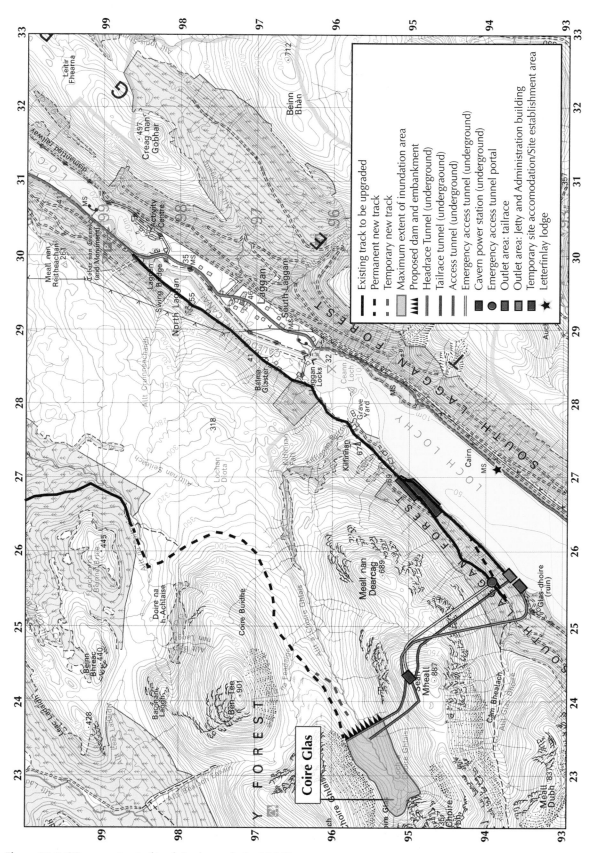

Figure 11.1 *OS map extract of Loch Lochy, scale 1 : 50 000*

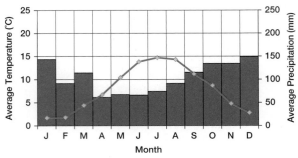

Figure 11.2: *Climate graph for Coire Glas and surrounding area*

Scottish Ministers recognise that the noise and visual impacts of the Development on local businesses and local residents, particularly during the construction period of the Development, are a major cause of concern.

Figure 11.3: *Letter issued by the Scottish Government's Energy Consents and Deployment Unit*

Figure 11.4: *Artist's rendering of proposed dam and reservoir*

Letterfinlay Lodge Hotel (located on the banks of Loch Lochy) owner Ian Smith had been planning a £2.5million expansion. He said:

'We are 45 seconds by speedboat from where the main works will be. They're talking about 25 lorries an hour on the A82 behind us. It's horrendous. I have planning permission for eight luxury apartments and a 50-berth marina. I can understand that Scotland needs a renewable energy policy but not at the expense of railroading through people's businesses. We employ up to 20 people. The SSE energy project when it's finished will employ 12.'

Figure 11.5: *View of local businessman*

Planning chief Malcolm MacLeod said:

'The development is expected to have impact on the local economy, both positive and negative. It is a significant initiative by a valued company with many assets in the Highlands, which has implications for the grid network and many other investments in renewable energy. There is potential for locals to gain from an investment of this size. The downside to the local economy is the adverse impact during the construction phase.'

Figure 11.6: *View of planning chief*

Answering the question

Read the conditions that need to be met for the location of the dam.

Read all the information carefully. Study the OS map extract.

Go through each point of the conditions and try to make a comment on it. These comments could be both positive and negative. The question asks you to evaluate so you should give your opinion using the evidence.

Starting with the rainfall, you could make the general point that, because this is a mountainous area, and rainfall tends to be greater in mountainous areas, there should be sufficient rainfall to fill the dam.

You can then refer to Figure 11.2: Climate graph for the area. It shows that there is rainfall all year round so a constant supply of water is available to fill the dam, so this is a suitable location. Although there is less rainfall in spring and the early summer months, there is no dry period so the dam should have a constant supply of water. Temperatures are not too high so less water will be lost through evaporation.

The dam will be an advantage to the farmers and the settlements along the valley floor as flooding, which could occur in times of high rainfall, will be prevented.

From the OS map it can be seen at grid reference 236956 there is evidence of a deep narrow valley which is suitable for damming as the dam length is reduced and there is a good storage area for the reservoir behind the dam. The area, as can be seen on the OS map, is remote. There is very little sign of human activity in the area so it will not disturb many people or cause people to be displaced by the new reservoir. There are no large settlements in the area, mostly single buildings along the edge of the loch; for example, the activity centre at 305983. There are tributaries in the area adding to the catchment area, for example at 236945. There is access to the area allowing the structure to be maintained.

Another part of the conditions is the visual impact. Local business people's viewpoints, as shown in Figure 11.5 as well as the letter from the government in Figure 11.3, indicate that there will be a visual impact. This area is popular with tourists as indicated on the map by footpaths, forest walks, the activity centre and the Letterfinlay Lodge Hotel. The information indicates that local businesses like the Letterfinlay Lodge Hotel feel that the dam structure might put people off visiting this well-known scenic area, thus adversely affecting their business. However, the artist's impression shows that the dam is below eye line and has a minimal impact. The area is full of lochs like Loch Lochy so the reservoir behind the dam fits in with the local scenery. This would therefore meet the condtions. Also, tourists might be attracted to visit the HEP station so it could increase local business as opposed to the area losing business.

The scheme, however, does not seem to meet the conditions as far as minimal disruption to the local people is concerned. The Scottish government in Figure 11.3 admits that there is going to be visual and noise impact, especially during construction. The main road runs past the hotel so the construction traffic will disturb the quiet beauty of the area and adversely affect the hotel business.

Overall, the conditions have been met. There is sufficient rainfall to fill the reservoir throughout the year. The glaciated area provides the deep narrow valley for the dam and other areas of water on the map indicate that the area has impermeable rock so water is retained on the surface. The map shows there are few settlements or farms in the area, so there will not be a lot of people displaced by the reservoir. Visual pollution is limited as the dam is hidden by the valley sides. However, noise pollution, especially during the construction period, could be a problem and cause local businesses like the hotel to lose customers. However, overall I feel the site of the dam meets the conditions.

Example question 2

Here is another scenario. Work through this on your own. Follow the advice given for scenario 1 to answer scenario 2. Remember, you need to evaluate the question and back up your answer with evidence. Peer mark it, then discuss your decisions with a partner.

A group of geography students, staying at the youth hostel in West Lulworth (832806), are planning to complete a circular walk along part of the South West Coast Path in Dorset. Working to the conditions below, a route has been proposed.

> **Conditions for Geography student field trip**
>
> The route should:
>
> - be suitable for all participants – one pupil has recently broken their leg and several pupils suffer with asthma
> - cause minimum environmental damage to the coastal area
> - have a suitable start/finish line
> - be scenic/interesting for participants.

Hint

Using four-/six-figure grid references as well as giving named examples will gain marks.

Hint

When answering this question make sure you refer to all of the sources provided in your answer. They have all been included for a reason, and this will help you pick up marks.

Hint

You should always include grid references from the OS map in your answer. If you are unsure of how to reference a point on an OS map, make sure you revise this carefully.

Study Figures 11.7–11.

Referring to map evidence and other information from the sources, **evaluate the suitability of the proposed route** in relation to the conditions for the circular walk around the Lulworth area. You should **suggest possible improvements** to the route.

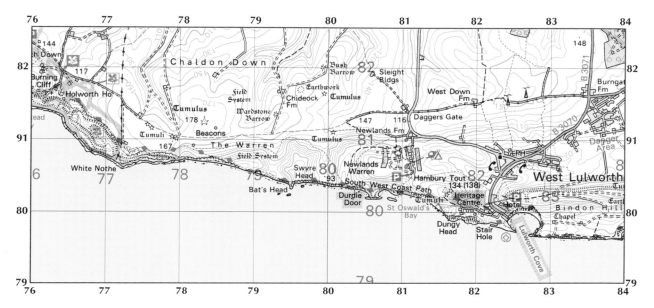

Figure 11.7: *OS map extract of Lulworth Cove, scale 1:50 000*

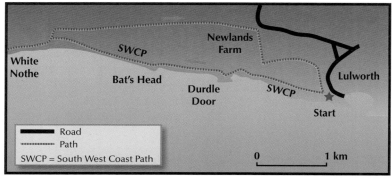

Figure 11.8: *Proposed route of circular walk*

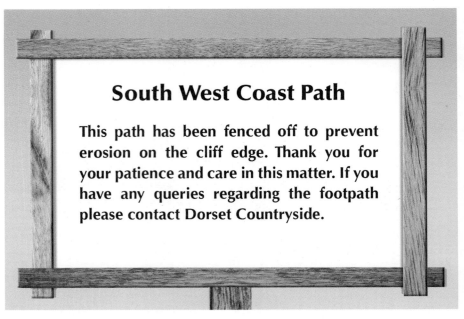

Figures 11.9 and 11.10: *Signposts along the South West Coast Path in Dorset*

THE DORSET NEWS

ALL ABOUT THE WORLD WE LIVE IN   *EXCLUSIVE NEWS TODAY*

NEW WARNINGS AFTER CLIFF FALL TRAGEDY

'Killer Cliffs of Dorset'

Warnings came today about the 'killer cliffs of Dorset' after yesterday's tragedy at Lulworth Cove when a landslip engulfed a school party in tons of rock and clay, killing a teacher and teenage pupil and seriously injuring two others. Unfortunately, these areas of scenic beauty are best avoided if possible in wet weather and with large parties.

Weather Alert

Today's weather

Temperatures will be cold this winter morning, with highs of 4°C. There will be heavy fog along the coastline; however this will burn away by the early afternoon. Sunny spells with frequent showers throughout the afternoon, leading to much cooler temperatures by this evening.

Figure 11.11: *Front page of local newspaper*

Text acknowledgements

Quotation by Charles E Kellogg © 1938, was taken from USDA Yearbook of Agriculture. Produced in the Public Domain.

Quotation from Peter Benton, Beyond 2011 Programme Director, was taken from the Office of National Statistics, used by permission.

Quotation from a speech by HRH The Prince of Wales to launch an Online Public Awareness Campaign to Save Rainforests, 5th May 2009, London, https://www.princeofwales.gov.uk/speech/speech-hrh-prince-wales-launch-online-public-awareness-campaign-save-rainforests-london

Quotation from UN Report: Our Common Future was produced by the United Nations.

Quotation from Rio Geographer - Filipe Bagatolli was taken from RioOnWatch https://www.rioonwatch.org/?p=11271. Used by permission.

Quotation from speech by William Hague given on 16th February 2006. was taken from: https://conservative-speeches.sayit.mysociety.org/speech/600128

Quotation from: Paul Wheelhouse, MSP and Former Minister for Environment and Climate Change is used by permission of Paul Wheelhouse.

Quotation from speech by Bam Ki-Moon, Former UN Secretary General is used by permission.

Quotation from the WWF is used by permission of the World Wildlife Fund for Animals.

Quotation from "Fury as Scottish councillors back controversial £800m hydro power scheme", Ian Smith, published in the Daily Record, 31 October 2012.

Quotation from "Fury as Scottish councillors back controversial £800m hydro power scheme", Malcolm MacLeod, published in the Daily Record, 31 October 2012.